Springer Undergraduate Mathematics Series

The Springer Undergraduate Mathematics Series (SUMS) is a series designed for undergraduates in mathematics and the sciences worldwide. From core foundational material to final year topics, SUMS books take a fresh and modern approach. Textual explanations are supported by a wealth of examples, problems and fully-worked solutions, with particular attention paid to universal areas of difficulty. These practical and concise texts are designed for a one- or two-semester course but the self-study approach makes them ideal for independent use.

More information about this series at https://link.springer.com/bookseries/3423

Robert Magnus

Metric Spaces

A Companion to Analysis

 Springer

Robert Magnus (iD)
Faculty of Physical Sciences
University of Iceland
Reykjavik, Iceland

ISSN 1615-2085   ISSN 2197-4144   (electronic)
Springer Undergraduate Mathematics Series
ISBN 978-3-030-94945-7   ISBN 978-3-030-94946-4   (eBook)
https://doi.org/10.1007/978-3-030-94946-4

Mathematics Subject Classification: 54E35, 54E45, 54E50, 54E52, 46B25

This Springer imprint is published by the registered company Springer Nature Switzerland AG
The registered company address is: Gewerbestrasse 11, 6330 Cham, Switzerland

To Edda and Friðrik

Preface

Open any book on multivariate[1] calculus, complex analysis, differential geometry, functional analysis, measure and integration, even mathematical physics or differential equations, from the last seventy years or so, and one is likely to see at the start some chapters introducing set theory and metric spaces. In order to pursue analysis beyond the level of single variable calculus, one must have a grounding in basic set theory and metric spaces. They are the handmaids of analysis.

Often metric spaces are taught as a part of a general topology course. Departing from this idea, the present work is an attempt to set out in a self-contained text the metric space theory that is needed for the various branches of analysis. The result is a handbook that will, it is hoped, prove useful for readers studying any of the subjects listed in the opening paragraph.

Originally the intention was to include set theory in the current work, preceding the treatment of metric spaces. However, the inordinate length of the resulting volume led to a change of plan. To compensate for the omission of set theory, a brief, unnumbered, preliminary chapter on sets has been included. This is, firstly, in order to list notations and concepts as used by the author, because they may differ from what may be found in commonly used textbooks, and secondly, in order to provide needed clarification.

One is apt to overlook the extraordinary simplification and clarity that concepts of set theory and metric spaces made possible in analysis. Beginning in 1799 Gauss, possibly the most brilliant mathematician of them all, gave several proofs of the Fundamental Theorem of Algebra. They all contained gaps. Nowadays it is a mere exercise to prove this theorem using the Bolzano-Weierstrass theorem, to the effect that every bounded sequence of complex numbers has a convergent subsequence. But Gauss lacked the theoretical machinery for this: firstly the set theory, needed to clarify the notion of sequence; secondly the metric space theory, needed to define the limit of a sequence of objects more general than real numbers; and thirdly an

---

[1] Throughout this text the term "multivariate" refers to studies involving functions of more than one variable. It does not refer to its more restrictive use in statistics.

accurate description of the properties of the real number field, needed to prove the Bolzano-Weierstrass theorem.

If this text were only a listing of definitions and theorems of metric spaces it would be unremittingly dry. Some significant excursions into analysis are presented to provide a sense of what can be accomplished with the material at hand. Such excursions encroach on material that would naturally appear in texts to which the current work is intended as a companion, and risk turning the present work into an incomplete tract on functional analysis. Therefore, in order to justify their inclusion, some attempt has been made to present impressive results, and to present each at the earliest point at which it is feasible.

These excursions include, for example, the Mazur-Ulam theorem (not always found in introductory books on functional analysis), the Stone-Weierstrass theorem, the theorem of Banach and Mazurkiewicz on nowhere differentiable functions, the Tietze extension theorem, space filling curves, and many other exciting and often challenging topics. They are optional reading and can be omitted without losing the main thread. For this reason they are marked with the symbol ($\Diamond$), as are exercises that refer back to material in such sections.

Exercises constitute an essential part of this text. Many introduce new notions and ask the reader to prove substantial results, which may be referred to at other points in the text. Outside the actual exercise sections there are also exercises embedded in the text; they urge the reader to engage immediately and constructively with the material. In some cases an embedded exercise may ask the reader to check or elucidate a point just made in the text.

The prerequisites for reading this text are a course in real analysis of one variable, including the real numbers, limits and continuity. This portion of knowledge is referred to throughout the text as fundamental analysis. A good grounding in set theory is needed; enough to enable the reader to handle sets in a safe and sound fashion. The topics needed from set theory are listed in the chapter on preliminaries. A familiarity with rigorous proofs is also desirable. However, this text is most likely to prove useful for advanced undergraduate or postgraduate students concurrently studying one or more of the topics listed in the opening paragraph.

Reykjavik, Iceland   Robert Magnus
October 2021

Contents

1 Metric Spaces 1
  1.1 Metrics 1
      1.1.1   Rationale for Metrics 1
      1.1.2   Defining Metric Space 2
      1.1.3   Exercises 5
  1.2 Examples of Metric Spaces 6
      1.2.1   Normed Spaces 6
      1.2.2   Subspaces 8
      1.2.3   Examples; Not Subspaces of Normed Spaces 9
      1.2.4   Pseudometrics 9
      1.2.5   Cauchy-Schwarz, Hölder, Minkowski 11
      1.2.6   Exercises 14
  1.3 Cantor's Middle Thirds Set 16
      1.3.1   Exercises 19
  1.4 The Normed Spaces of Functional Analysis 20
      1.4.1   Sequence Spaces 20
      1.4.2   Function Spaces 22
      1.4.3   Spaces of Continuous Functions 23
      1.4.4   Spaces of Integrable Functions 23
      1.4.5   Hölder's and Minkowski's Inequalities for Integrals 24
      1.4.6   Exercises 25

2 Basic Theory of Metric Spaces 29
  2.1 Balls in a Metric Space 29
      2.1.1   Limit of a Convergent Sequence 30
      2.1.2   Uniqueness of the Limit 31
      2.1.3   Neighbourhoods 32
      2.1.4   Bounded Sets 32
      2.1.5   Completeness; a Key Concept 32
      2.1.6   Exercises 35

2.2   Open Sets, and Closed   37
        2.2.1   Open Sets   38
        2.2.2   Union and Intersection of Open Sets   39
        2.2.3   Closed Sets   39
        2.2.4   Union and Intersection of Closed Sets   40
        2.2.5   Characterisation of Open and Closed Sets by Sequences ...   41
        2.2.6   Interior, Closure and Boundary   42
        2.2.7   Limit Points of Sets   43
        2.2.8   Characterisation of Closure by Limit Points   43
        2.2.9   Subspaces   45
        2.2.10   Open and Closed Sets in a Subspace   45
        2.2.11   Exercises   47
2.3   Continuous Mappings   52
        2.3.1   Defining Continuity   52
        2.3.2   New Views of Continuity   53
        2.3.3   Limits of Functions   55
        2.3.4   Characterising Continuity by Sequences   57
        2.3.5   Lipschitz Mappings   57
        2.3.6   Examples of Continuous Functions   58
        2.3.7   Exercises   59
2.4   Continuity of Linear Mappings   63
        2.4.1   Continuity Criterion   64
        2.4.2   Operator Norms   66
        2.4.3   Exercises   68
2.5   Homeomorphisms and Topological Properties   70
        2.5.1   Equivalent Metrics   72
        2.5.2   Exercises   73
2.6   Topologies and σ-Algebras   74
        2.6.1   Order Topologies   77
        2.6.2   Exercises   79
        2.6.3   Pointers to Further Study   82
2.7   ($\Diamond$) Mazur-Ulam   82
        2.7.1   Exercises   84

3   Completeness of the Classical Spaces   87
3.1   Coordinate Spaces and Normed Sequence Spaces   87
        3.1.1   Completeness of $\mathbb{R}^n$ 87
        3.1.2   Completeness of ℓ^p   88
        3.1.3   Exercises   90
3.2   Product Spaces   90
        3.2.1   Finitely Many Factors   91
        3.2.2   Infinitely Many Factors   91
        3.2.3   The Space $2^{\mathbb{N}_+}$ and the Cantor Set   93
        3.2.4   Subspaces of Complete Spaces   95
        3.2.5   Exercises   96

3.3   Spaces of Continuous Functions 99
     3.3.1   Uniform Convergence 99
     3.3.2   Series in Normed Spaces 101
     3.3.3   The Weierstrass M-Test 103
     3.3.4   The Spaces $C(\mathbb{R})$ and $C^p(\mathbb{R})$ 103
     3.3.5   Exercises 105
3.4   ($\lozenge$) Rearrangements 108
     3.4.1   Vector Series 108
     3.4.2   Exercises 111
     3.4.3   Pointers to Further Study 111
3.5   ($\lozenge$) Invertible Operators 112
     3.5.1   Fredholm Integral Equation 115
     3.5.2   Exercises 116
     3.5.3   Pointers to Further Study 118
3.6   ($\lozenge$) Tietze 118
     3.6.1   Formulas for an Extension 121
     3.6.2   Exercises 121
     3.6.3   Pointers to Further Study 123

4 Compact Spaces 125
4.1   Sequentially Compact Spaces 125
     4.1.1   Continuous Functions on Sequentially Compact Spaces 126
     4.1.2   Bolzano-Weierstrass in $\mathbb{R}^n$ 126
     4.1.3   Sequentially Compact Sets in $\mathbb{R}^n$ 127
     4.1.4   Sequentially Compact Sets in Other Spaces 127
     4.1.5   The Space $C(M)$ 128
     4.1.6   Exercises 129
4.2   The Correct Definition of Compactness 130
     4.2.1   Thoughts About the Definition 131
     4.2.2   Compact Spaces and Compact Sets 132
     4.2.3   Continuous Functions on Compact Spaces 134
     4.2.4   Uniform Continuity 135
     4.2.5   Exercises 136
4.3   Equivalence of Compactness and Sequential Compactness 137
     4.3.1   Relative Compactness 141
     4.3.2   Local Compactness 142
     4.3.3   Exercises 143
4.4   Finite Dimensional Normed Vector Spaces 150
     4.4.1   Exercises 153
4.5   ($\lozenge$) Ascoli 154
     4.5.1   Peano's Existence Theorem 157
     4.5.2   Exercises 160
     4.5.3   Pointers to Further Study 162

5   Separable Spaces 163
   5.1   Dense Subsets of a Metric Space 163
       5.1.1   Defining a Vector-Valued Integral 165
       5.1.2   Exercises 168
   5.2   Separability 171
       5.2.1   Second Countability 172
       5.2.2   Exercises 173
   5.3   (◊) Weierstrass 176
       5.3.1   Exercises 180
       5.3.2   Pointers to Further Study 182
   5.4   (◊) Stone-Weierstrass 182
       5.4.1   Exercises 185
       5.4.2   Pointers to Further Study 186

6   Properties of Complete Spaces 187
   6.1   Cantor's Nested Intersection Theorem 187
       6.1.1   Categories 188
       6.1.2   Exercises 190
   6.2   (◊) Genericity 193
       6.2.1   Exercises 196
       6.2.2   Pointers to Further Study 199
   6.3   (◊) Nowhere Differentiability 199
       6.3.1   Exercises 202
       6.3.2   Pointers to Further Study 203
   6.4   Fixed Points 203
       6.4.1   Exercises 204
   6.5   (◊) Picard 206
       6.5.1   Exercises 209
   6.6   (◊) Zeros 209
       6.6.1   Exercises 213
       6.6.2   Pointers to Further Study 216
   6.7   Completion of a Metric Space 216
       6.7.1   Other Ways to Complete a Metric Space 218
       6.7.2   Exercises 218

7   Connected Spaces 219
   7.1   Connectedness 219
       7.1.1   Connected Sets 221
       7.1.2   Rules for Connected Sets 221
       7.1.3   Connected Subsets of $\mathbb{R}$ 223
       7.1.4   Exercises 223
   7.2   Continuous Mappings and Connectedness 224
       7.2.1   Continuous Curves 224
       7.2.2   Arcwise Connectedness 225
       7.2.3   Exiting a Set 226
       7.2.4   Exercises 227

7.3   Connected Components.. 229
      7.3.1   Examples of Connected Components 229
      7.3.2   Arcwise Connected Components 232
      7.3.3   Exercises 232
7.4   (◊) Peano... 236
      7.4.1   Exercises 236
      7.4.2   Pointers to Further Study 237

Afterword 239
Index... 241

Preliminaries on Sets

We list here certain conventions, notations and ideas concerning set theory which are assumed throughout this text. Our purpose is to clarify, but not necessarily to define, notions used in this text that should already be familiar to the reader, especially where these may differ notationally or in substance from those used by other authors. We do not list all the propositions or concepts of set theory that may be referred to in the text.

Basic Relations

These are set membership, $x \in A$ (x is an element of the set A) and set inclusion, $A \subset B$ (A is a subset of B, or B includes A). The latter allows for equality of A and B.

Basic Operations

The binary operations are $A \cup B$ (union), $A \cap B$ (intersection), $A \setminus B$ (set difference) and $A \mathbin{\triangle} B$ (symmetric difference). There is a unary operation, A^c (complement). The cartesian product of sets A and B is written $A \times B$.

Writing Predicates

We use logical connectives and quantifiers to write predicates. For example, we can define a predicate $\psi(x)$, where x ranges over all positive integers, to mean

$$x \geq 2 \wedge (\forall y)((y \in \mathbb{N}_+ \wedge y \mid x) \Rightarrow y = 1 \vee y = x)$$

The predicate $\psi(x)$ says that x is a prime. Here $\mathbb{N}_+$ denotes the set of positive integers and the relation $y \mid x$ says that y divides x. We use the logical connectives $\wedge$ (conjunction or *logical and*), $\vee$ (disjunction or *logical or*) and $\Rightarrow$ (logical implication). We also use the universal quantifier $\forall y$ asserting that the statement following it holds for all y. As another example, the predicate $\phi(x)$ could mean

$$(\exists y)(y \in \mathbb{N}_+ \wedge x = y^2)$$

The predicate $\phi(x)$ says that x is a square. Here we have the existential quantifier $\exists y$ asserting that y exists for which the statement following it is true.

Set-Building Rules

1. We form the set containing given elements, usually finitely many, by enclosing them in curly brackets. For example

$$\{1, 2\}, \quad \{\{1, 2\}, 3\}$$

2. Given a set E and a predicate $\psi(x)$, we form the set A of all elements x of E for which $\psi(x)$ is true, by writing

$$A = \{x \in E : \psi(x)\}$$

For example, using the predicate $\psi(x)$, that says that x is a prime, we form the set of all prime numbers

$$A = \{x \in \mathbb{N}_+ : \psi(x)\}.$$

3. Given a family $(A_i)_{i \in I}$ of sets we can form the union

$$\bigcup_{i \in I} A_i$$

The defining property of the union is that $x \in \bigcup_{i \in I} A_i$ if and only if there exists $i \in I$, such $x \in A_i$.

4. Given a family $(A_i)_{i \in I}$ of sets we can form the intersection

$$\bigcap_{i \in I} A_i$$

The defining property of the intersection is that $x \in \bigcap_{i \in I} A_i$ if and only, for all $i \in I$, we have $x \in A_i$.

5. Given a set X we can form the power set $\mathcal{P}(X)$, intuitively, the set of all subsets of X. The defining property of $\mathcal{P}(X)$ is that $A \in \mathcal{P}(X)$ if and only if $A \subset X$.

6. (Axiom of choice) Given a family $(A_i)_{i \in I}$ of sets, all of which are non-empty, there exists a family $(a_i)_{i \in I}$ (with the same index set), such that $a_i \in A_i$ for all i. This is a set-building rule because a family, like a function, can be seen as a certain set—a fact that need not concern us further.

7. Given a family of $(A_i)_{i \in I}$ of sets we can form the product

$$\prod_{i \in I} A_i$$

This is the set of all families $(a_i)_{i \in I}$ with $a_i \in A_i$. If all the sets A_i are the same, say $A_i = X$, we can write the product using exponentiation as X^I.

8. We already have at our disposal a number of sets. The empty set $\emptyset$ has no elements. The natural numbers, taken to include 0, form a set denoted by $\mathbb{N}$. This has the defining property that it is the minimum set (in the sense of inclusion) that contains the natural number 0, and for all x, if x is a natural number and $x \in \mathbb{N}$ then $x + 1 \in \mathbb{N}$. For obvious reasons the set $\mathbb{N}$ is infinite. The other sets of analysis, $\mathbb{Z}$ (the signed integers), $\mathbb{Q}$ (the rationals), $\mathbb{R}$ (the real numbers) and $\mathbb{C}$ (the complex numbers) can be built from $\mathbb{N}$ using the foregoing set-building rules.

A comment on union and intersection. Although we usually write the union and intersection of arbitrarily large collections of sets by presenting the sets as the terms of a family, these operations can, and perhaps should, be viewed more formally as *unary operations on sets whose elements are sets*. Thus if X is a set, whose elements are all sets, it can be convenient to write

$$\bigcup X \quad \text{and} \quad \bigcap X$$

to denote the union of all the sets that are members of X, and the intersection of all the sets that are members of X, respectively.

Relations and Functions

The reader should be familiar with the notion of a relation with domain X and codomain Y (where X and Y are sets). Here we may have an order relation (for which $X = Y$), possibly of a specialised type, such as a total ordering or a well-ordering. Equivalence relations are important, with the associated partition of the set X into equivalence classes.

As usual we write $f : X \to Y$ to mean that f is a function (or mapping; it means the same) with domain X and codomain Y. Associated concepts of injectivity, surjectivity, bijectivity and inverse function should be familiar. We stick to the convention that a bijective mapping is both injective and surjective. So whether an injective mapping is bijective depends on the codomain. The composition of functions is written $f \circ g$; the understanding being "apply g then f".

A family $(a_i)_{i \in I}$ of elements a_i from a set A, with index set I is really just a function $f : I \to A$, but presented differently; the meaning is that $a_i = f(i)$. The use of parentheses is important and emphasises that terms of the family are not necessarily distinct; we can have $a_i = a_j$ though $i \neq j$. We can form the *set of all terms* of the family by switching to curly brackets:

$$\{a_i : i \in I\}.$$

For example, we can form a family (an infinite sequence) $(a_n)_{n \in \mathbb{N}}$ by setting $a_n = 1$ for all n. Then $\{a_n : n \in \mathbb{N}\} = \{1\}$. An infinite sequence is a family whose index set is $\mathbb{N}$ or $\mathbb{N}_+$. The presentation $(a_n)_{n=1}^{\infty}$ is usually used.

Cardinals

The cardinal number of $\mathbb{N}$ is denoted by $\aleph_0$. The cardinal number of $\mathbb{R}$ is denoted by $\mathbf{c}$. Occasionally (but only for some exercises) some very simple cardinal arithmetic may be needed. The cardinal of a set X is sometimes denoted by $|X|$. The set of all functions with domain X and codomain Y is denoted by Y^X. Its cardinal is the exponential $|Y|^{|X|}$. A set is countable when it is finite or, if infinite, it has the cardinal $\aleph_0$. Otherwise it is uncountable.

Other Notions

Partitions A partition of a set X is a family $(A_i)_{i \in I}$ of non-empty subsets of X, such that $A_i \cap A_j = \emptyset$ whenever $i \neq j$, and $\cup_{i \in I} A_i = X$. Often, viewing a partition as a family is unnatural; instead we can say that a partition of X is a subset $D \subset \mathcal{P}(X) \setminus \{\emptyset\}$, the elements of which are pairwise disjoint sets, and satisfying $\bigcup D = X$.

Zorn's Lemma Some exercises may require this fundamental existence theorem, which is equivalent to the axiom of choice. It asserts that if X is a partially ordered set for which every chain (subset that is totally ordered) has an upper bound, then X contains a maximal element.

Restriction Given a mapping $f : X \to Y$ and a subset $A \subset X$, the restriction of f to A is denoted by $f|_A$.

Exponentiation Given two sets X and Y the exponential Y^X is the set of all mappings with domain X and codomain Y. In particular, $X^{\mathbb{N}_+}$ is the set of all sequences $(x_n)_{n=1}^{\infty}$ with terms in X.

Notation for Intervals We use Bourbaki's (or the European) notation for intervals. Thus the open interval with endpoints a and b (where $a < b$) is written $]a, b[$, whilst the closed interval is written $[a, b]$. The notation (a, b) is used only for an ordered pair. The set of integers k in the range $m \le k \le n$ is denoted by $[[m, n]]$.

Chapter 1
Metric Spaces

Metric system, international decimal system of weights and measures, based on the metre for length and the kilogram for mass, that was adopted in France in 1795 and is now used officially in almost all countries.

Definition of metric system from Encyclopaedia Britannica

1.1 Metrics

This purpose of this chapter is twofold: to motivate and define the notion of metrics and metric spaces; and to show that examples of metric spaces abound in the various areas of analysis. The actual theory of metric spaces begins in chapter 2.

1.1.1 Rationale for Metrics

We assume that the reader has a knowledge of fundamental analysis, by which we mean a rigorous account of the analysis of functions of one real variable. It is grounded on the notion that the real numbers $\mathbb{R}$ constitute a Dedekind complete, ordered field. The ordering enables one to define the limit of a sequence of real numbers, and the continuity of a function of a real variable having real values. Together with completeness it suffices to develop differential and integral calculus in a logically satisfactory fashion.

We recall the definition of the limit of a sequence $(a_n)_{n=1}^{\infty}$ of real numbers. We say that the sequence is convergent and has the limit t (where t is a real number), and we write $\lim_{n\to\infty} a_n = t$, when the following condition is satisfied:

For all $\varepsilon > 0$ there exists a natural number N, such that $|a_n - t| < \varepsilon$ for all $n \geq N$.

© The Author(s), under exclusive license to Springer Nature Switzerland AG 2022
R. Magnus, *Metric Spaces*, Springer Undergraduate Mathematics Series,
https://doi.org/10.1007/978-3-030-94946-4_1

This complicated assertion requires three nested quantifiers. In slightly simplified logical notation the definition of $\lim_{n\to\infty} a_n = t$ is:

$$(\forall \varepsilon \in \mathbb{R}_+)(\exists N \in \mathbb{N})(\forall n \in \mathbb{N})(n \geq N \Rightarrow |a_n - t| < \varepsilon)$$

We may also note that the definition makes sense in any ordered field, because one may define the absolute value $|a|$ to be $\max(a, -a)$.

Now consider a sequence of complex numbers, which we shall also denote by $(a_n)_{n=1}^{\infty}$, although the use of 'z' instead of 'a' might be more common as a visual cue that complex numbers are under consideration. We also suppose t to be a complex number. We can reinterpret the logical definition of convergence, as just given, by taking $|a_n - t|$ to be the *modulus of the complex number $a_n - t$*.

We can go further, leaving complex numbers behind, and suppose that the elements a_n and t are vectors in $\mathbb{R}^m$ and reinterpret $|a_n - t|$ as *the Euclidean length of the vector $a_n - t$*. In these extensions of limit, we are departing completely from operating in an ordered field. Neither the complex plane, nor the space of coordinate vectors $\mathbb{R}^m$, is an ordered field.

In the fashion just described we can define the notions of convergence of a sequence of complex numbers, and convergence of a sequence of vectors in $\mathbb{R}^m$, using the same logical sentence as was used for real number sequences, but reinterpreting the expression $|a_n - t|$. In each of these cases it is natural to think of $|a_n - t|$ as the *distance between the points a_n and t*.

It would appear from this that we can define the limit of a *convergent sequence of elements in an arbitrary set X*, if we have a notion of distance between pairs of elements of X. Such a notion of distance is called a metric on X.

Exercise Read carefully the definition of limit given in the first paragraph of this chapter. The wording is idiomatically correct English but could be thought ambiguous on the grounds that the intended meaning could be

$$(\forall n \in \mathbb{N})(\forall \varepsilon \in \mathbb{R}_+)(\exists N \in \mathbb{N})(n \geq N \Rightarrow |a_n - t| < \varepsilon).$$

What does this incorrect definition assert about the sequence? Note that two consecutive universal quantifiers '$(\forall a)(\forall b)$' are equivalent to 'for all a and b'.

1.1.2 Defining Metric Space

Let X be a set. We wish to assign to every pair of points x and y in X a real number which we think of as the distance between the points. Note that we call the elements of X points; this is customary, and helps to activate geometric intuition. What properties should this notion of "distance" possess?

Obviously it should not be negative. It appears that three properties in addition to this are enough to engender a remarkable body of useful theory. The first two are obvious enough; the third injects a modicum of geometry into the works.

Definition A metric on the set X is a function $d : X \times X \to [0, \infty[$ that satisfies the following conditions (or axioms for metrics):

1) For all x and y in X we have $d(x, y) = 0$ if and only if $x = y$.
2) (Symmetry) For all x and y in X we have $d(x, y) = d(y, x)$.
3) (Triangle inequality) For all x, y and z in X we have

$$d(x, z) \leq d(x, y) + d(y, z).$$

Definition A metric space is a set X equipped with a metric; more precisely it is an ordered pair (X, d), where X is a set and d a metric on X.

We will often say "Let X be a metric space with metric d" to mean that (X, d) is a metric space. Occasionally it is convenient to allow ∞ as a possible value for $d(x, y)$. The triangle inequality is then interpreted in the obvious fashion, taking $a + \infty = \infty$ and $a \leq \infty$ for all $a \in \mathbb{R} \cup \{\infty\}$. Allowing ∞ makes subtraction tricky (as in the following proposition); so usually we assume, unless stated otherwise, that $d(x, y)$ is finite.

Proposition 1.1 *Let X be a metric space with metric d. For all x, y and z in X we have*

$$|d(x, z) - d(y, z)| \leq d(x, y).$$

Proof We have

$$d(x, z) \leq d(x, y) + d(y, z)$$

by the triangle inequality, so that

$$d(x, z) - d(y, z) \leq d(x, y).$$

Now interchange x and y. Using the symmetry of the metric d (property 2 of metrics) we also obtain

$$d(y, z) - d(x, z) \leq d(x, y).$$

By the last two inequalities we find

$$|d(x, z) - d(y, z)| \leq d(x, y).$$

$\square$

Two of the simplest things we can do with a metric are defining the diameter diam A of a non-empty subset A of X, and the distance $d(x, A)$ from a point x to the non-empty subset A. These are as follows:

$$\operatorname{diam} A := \sup\{d(x, y) : x \in A,\ y \in A\},$$

$$d(x, A) := \inf\{d(x, y) : y \in A\}.$$

Note that we allow $\operatorname{diam} A = \infty$ if the set of numbers for which the supremum is required is unbounded. Note also that whilst $d(x, A) = 0$ if $x \in A$, as should be clear from the definition, it is possible that $d(x, A) = 0$ though x is outside A (see exercise 2). We exclude the empty set from these definitions as we don't get anything very sensible (just try to calculate $d(x, \emptyset)$ and $\operatorname{diam} \emptyset$!)

The following property of $d(x, A)$, which extends proposition 1.1, is often useful. It is a nice application of the notion of infimum (greatest lower bound) and the triangle inequality.

Proposition 1.2 *Let (X, d) be a metric space, let $x \in X$, $y \in X$ and $A \subset X$, with A non-empty. Then*

$$|d(x, A) - d(y, A)| \le d(x, y).$$

Proof Let $\varepsilon > 0$. By the definition of $d(x, A)$ as an infimum there exists $z \in A$ such that

$$d(x, z) < d(x, A) + \varepsilon.$$

Furthermore

$$d(x, z) + d(x, y) \ge d(y, z) \ge d(y, A).$$

Hence

$$d(x, A) + d(x, y) + \varepsilon > d(y, A)$$

that is,

$$d(y, A) - d(x, A) < d(x, y) + \varepsilon.$$

This holds for all $\varepsilon > 0$, whilst z (the only thing that depended on ε) no longer appears. We conclude that

$$d(y, A) - d(x, A) \le d(x, y).$$

Finally we interchange x and y and obtain the desired conclusion.   □

When a metric has been defined on a set X we can define the notion of limit of a sequence of elements of X. To give meaning to the expression '$\lim_{k \to \infty} x_k = p$', we only have to replace the expression '$|x_n - p|$', as would be used if the variables were real numbers, by '$d(x_n, p)$', in the formal definition of limit.

Definition Let (X, d) be a metric space and let $(x_k)_{k=1}^{\infty}$ be a sequence in X. Then the sequence $(x_k)_{k=1}^{\infty}$ is said to converge to a point p in X if the following condition is satisfied:

For all $\varepsilon > 0$ there exists $N \in \mathbb{N}$, such that $d(x_k, p) < \varepsilon$ for all $k \geq N$.

As usual, if $(x_k)_{k=1}^{\infty}$ converges to p we write

$$\lim_{k \to \infty} x_k = p.$$

The main study of limits will be taken up in detail in chapter 2, but we state a definition here to give some preliminary justification for the study of metric spaces. We will replace this definition with another in chapter 2; though it is equivalent to this one it is somewhat nicer.

The remainder of this chapter will be taken up with providing justification of a different sort: that there are plenty of examples of metric spaces that arise in ordinary mathematics, particularly in analysis.

1.1.3 Exercises

1. Let (X, d) be a metric space, and let A and B be subsets of X. Suppose that $A \cap B \neq \emptyset$. Show that

$$\text{diam}\,(A \cup B) \leq \text{diam}\,(A) + \text{diam}\,(B).$$

   Show, by an example, that this does not necessarily hold if $A \cap B = \emptyset$.
2. Find an example of a subset A of $\mathbb{R}$, and a point $p \notin A$, such that $d(p, A) = 0$.
3. Let (X, d) be a metric space. A subset A of X is said to be bounded if there exists a point $a \in X$ and $R > 0$, such that for all $x \in A$ we have $d(a, x) < R$.

   a) Fix a point $b \in X$. Show that a set A is bounded if and only if there exists $R > 0$, such that for all $x \in A$ we have $d(b, x) < R$. The point here is that the same point b can be used to test sets for boundedness, independent of the set in question.
   b) Show that the union of finitely many bounded sets is bounded.
   c) Show that a non-empty set is bounded if and only if its diameter is finite.

4. Let d be a metric on the set X. Let $d'(x, y) = \min(d(x, y), 1)$. Show that d' is also a metric on X.
5. A more general version of the previous exercise. Let $f : [0, \infty[\to [0, \infty[$ and suppose that f satisfies a weak concavity condition: if $0 \leq s \leq t$ and $0 \leq \lambda \leq 1$,

then

$$f\big(\lambda s + (1 - \lambda)t\big) \geq \lambda f(s) + (1 - \lambda)f(t).$$

Suppose also that $f(0) = 0$ and that f is not identically 0.

a) Show that f is continuous in $]0, \infty[$, increasing (not necessarily strictly), and
   satisfies $f(t) = 0$ if and only if $t = 0$.
b) Show that f satisfies $f(s + t) \leq f(s) + f(t)$ for all $s > 0, t > 0$.
c) Let d be a metric on a set X. Show that $f \circ d$ is also a metric on X.

Hint. The reader should recall some facts from analysis about concave functions, for example, that they are continuous in any open interval; or that through any point in the graph of a concave function there is a line that does not pass below the graph. Important examples that are encountered in the context of metrics are $f(x) = x/(1 + x)$ and $f(x) = \min(x, 1)$.

6. The defining property for the relation '$\lim_{k \to \infty} x_k = p$' can be rewritten so that ε no longer appears, being based instead on the limit of a numerical sequence. Let (X, d) be a metric space, let $(x_k)_{k=1}^{\infty}$ be a sequence in X and let p a point in X. Prove that $\lim_{k \to \infty} x_k = p$ if and only if $\lim_{k \to \infty} d(x_k, p) = 0$.

1.2   Examples of Metric Spaces

1.2.1   Normed Spaces

The most important examples of metrics in multivariate calculus are furnished by the spaces of real coordinate vectors $\mathbb{R}^n$, as listed here:

a) $\mathbb{R}$ with $d(x, y) = |x - y|$.

Here $|x|$ denotes the absolute value of x. We will use this as the default metric on $\mathbb{R}$, sometimes calling it *the usual metric*.

b) $\mathbb{R}^n$ with $d(x, y) = |x - y|$.

Here $|x|$ denotes the Euclidean length of the vector $x = (x_1, x_2, \ldots, x_n)$:

$$|x| = \sqrt{x_1^2 + x_2^2 + \cdots + x_n^2}.$$

c) $\mathbb{R}^n$ with $d(x, y) = \|x - y\|_1$.

Here

$$\|x\|_1 := |x_1| + |x_2| + \cdots + |x_n|.$$

This is a special case of the next item.

d) $\mathbb{R}^n$ with $d(x, y) = \|x - y\|_p$.

Here p is a fixed real number, such that $p \geq 1$, and

$$\|x\|_p := \left(|x_1|^p + |x_2|^p + \cdots + |x_n|^p\right)^{\frac{1}{p}}.$$

e) $\mathbb{R}^n$ with $d(x, y) = \|x - y\|_\infty$. Here

$$\|x\|_\infty := \max \left(|x_1|, |x_2|, \ldots, |x_n|\right).$$

These examples should already indicate the ubiquity of metrics in multivariate analysis. That they satisfy the axioms for metrics will be shown shortly; only the triangle inequality is not obvious.

In all these examples the set is a vector space over $\mathbb{R}$ and the metric satisfies $d(x, y) = d(x - y, 0)$. The distance between x and y is the same as the distance from $x - y$ to the point 0, and depends only on the vector $x - y$. In fact $d(x, y)$ may be thought of as a way to assign a length to the vector $x - y$. This can be generalised and leads to an immensely important notion.

Definition Let E be a vector space over the real field $\mathbb{R}$. A norm on E is a function with domain E and codomain $[0, \infty[$, its value at a vector x usually indicated by $\|x\|$ (or something similar), that satisfies the following properties:

1) $\|x\| = 0$ if and only if $x = 0$.
2) $\|\lambda x\| = |\lambda| \, \|x\|$ for all $x \in E$ and $\lambda \in \mathbb{R}$.
3) (Triangle inequality) $\|x + y\| \leq \|x\| + \|y\|$ for all x and y in E.

A normed vector space, or simply, a normed space, is a vector space over $\mathbb{R}$ that is equipped with a norm. The quantities $|x| = \|x\|_2$ (Euclidean length), $\|x\|_p$ and $\|x\|_\infty$ are distinct norms that can be defined on the vector space $\mathbb{R}^n$. A vector in $\mathbb{R}^n$ acquires different lengths according to which norm is in use. It is misleading to say that vectors are quantities that have direction and length, as is taught in school mathematics. In order to compare the lengths of two vectors in $\mathbb{R}^n$ we need a norm, and there is a multitude of possible norms. Only vectors that are parallel can be compared without a norm.

We continue the list of commonly occurring metric spaces with:

f) E with the metric $d(x, y) := \|x - y\|$, where E is a vector space over $\mathbb{R}$ equipped with a norm $\| \cdot \|$.

We shall shortly look at some important examples of these spaces where E is not a coordinate space such as $\mathbb{R}^n$. The subject of functional analysis is, broadly, the study of these spaces in cases when E is an infinite-dimensional vector space, typically a vector space of functions. This may partly explain the term 'functional', making it an adjective (as in 'electrical engineer'). However, another theory is that it is a noun, there being objects called functionals that are important in linear algebra. Functional analysis is then an abstract theory for the study of functionals. Naturally it is much more than this and one may perhaps be forgiven for thinking that actually it means quite literally "analysis that works".

We framed the definition of normed vector space for a vector space over the real field $\mathbb{R}$. Just as important are normed vector spaces over the complex field $\mathbb{C}$. No change in the definition is needed except that $|\lambda|$ should be understood as the modulus of the complex number λ. The spaces $\mathbb{C}^n$ are the simplest examples, with the range of norms

$$\|a\|_p = \Big(\sum_{k=1}^{n} |a_k|^p \Big)^{1/p}, \quad (p \geq 1)$$

and

$$\|a\|_\infty = \max_{1 \leq k \leq n} |a_k|$$

where, again, we must interpret $|a_k|$ as the modulus of the complex number a_k.

1.2.2   Subspaces

The set of examples of metric spaces is vastly increased by a simple observation. Let (X, d) be a metric space and let $A \subset X$. Then A may be regarded as a metric space in its own right, by restricting the metric d to pairs of points in A. The metric space (A, d) is called a *subspace of* (X, d). If, without mentioning the metric, we say that A is a subspace of X, we mean just this: that A inherits the metric of X, by restriction, and is to be viewed as a metric space in its own right.

Now any subset of $\mathbb{R}^n$ can be viewed as a metric space, by restricting the metric of $\mathbb{R}^n$ (assuming we agree as to what this is). This shows the great strength of the metric concept. Any set studied in geometry is a metric space. But it also reveals a curious weakness. A circle T in the plane $\mathbb{R}^2$ is a metric space. We can give $\mathbb{R}^2$ the Euclidean norm and regard T as a subspace. The distance between two points in T is then the Euclidean length of the chord joining them. But if we want to see T as a metric space in its own right, would it not be more natural to regard the distance as the length of the shorter of the two arcs joining them? Then we don't have to go outside T to define the distance.

Similarly for a sphere regarded as a subspace of $\mathbb{R}^3$. Of course it all depends on context whether the distance between the North Pole and the South Pole should be the line segment joining them, or whether the great circle distance is meant. Our modern metric system intended originally to define the metre so that the great circle distance from the North Pole to the equator should be 10,000,000 metres. Actually, modern measurements give it as nearly 2,000 metres more.

We will see that some of the most important properties of spaces that one studies using metrics do not depend so much on the actual metric, but more so on the fact that some metric is available. These include the so-called topological properties, which we will encounter later.

1.2.3 Examples; Not Subspaces of Normed Spaces

We continue the list of examples of metric spaces with some that are not normed spaces, nor are they subspaces of normed spaces.

g) Any set X can be equipped with a metric. For each x and y in X we set $d(x, y) = 0$ if $x = y$ and $d(x, y) = 1$ if $x \neq y$.

h) $X = \mathbb{Z}$ with the dyadic metric.

This is defined as follows. If $x \neq y$ we set $d(x, y) = 2^{-v}$ where 2^v is the highest power of 2 that divides $x - y$. We set $d(x, x) = 0$. For example $d(2, 3) = 1$, $d(2, 4) = \frac{1}{2}$. In this funny metric 1 is much closer to 1025 than is 1024. Any prime p can replace 2 and the resulting metric on $\mathbb{Z}$ is called the p-adic metric (when p is unspecified—triadic, 5-adic, 7-adic,... in particular cases).

i) X is a graph or net with the path length metric.

For any two vertices x and y we let $d(x, y)$ be the number of edges in the shortest path joining x and y. If there is no path joining x and y we let $d(x, y) = \infty$. This is one of those cases when it seems natural to admit ∞ as a possible distance.

1.2.4 Pseudometrics

One sometimes encounters a function $d : X \times X \to [0, \infty[$ that satisfies axioms 2 and 3 for metrics, but not axiom 1 in its entirety. It is called a pseudometric when $d(x, x) = 0$ for all x, but there exist x and y, such that $x \neq y$ and $d(x, y) = 0$.

A pseudometric can be used to obtain a metric space by a natural procedure. To follow it, you will need to know about equivalence relations and equivalence classes.

Exercise Let X be a set equipped with a pseudometric d. Define the relation: $x \equiv y$ if and only if $d(x, y) = 0$.

a) Show that $x \equiv y$ is an equivalence relation on X.
b) Denote the equivalence class of x by $[x]$. Show that the set of equivalence classes may be made into a metric space, and a metric D defined on it, by setting

$$D([x], [y]) = d(x, y).$$

Hint. You will need to show that D is well-defined; it does not depend on the choice of representative elements of the equivalence classes.

A nice example of a pseudometric arises when we try to define a metric on the power set $\mathcal{P}(X)$, using a metric d defined on X. The outcome is the Hausdorff metric (or pseudometric). This can be used to define the distance between two subsets of the plane, and has applications to building pattern recognition programs in artificial intelligence.

Let (X, d) be a metric space. Recall the definition of $d(x, A)$, the distance from a point x to a non-empty set A, defined by

$$d(x, A) = \inf\{d(x, y) : y \in A\}.$$

For a pair, A and B, of non-empty subsets of X, we define

$$d_H(A, B) = \max\left(\sup_{x \in A} d(x, B),\ \sup_{y \in B} d(y, A)\right).$$

The value ∞ has to be allowed if we require the supremum of an unbounded set of numbers. The empty set is explicitly excluded. In general d_H is only a pseudometric.

Exercise Show that d_H is a pseudometric and not a metric, by computing $d_H(A, B)$, where $X = \mathbb{R}$, $A = [0, 1]$, $B =]0, 1[$.

Proof that d_H Satisfies the Triangle Inequality. For all pairs of subsets, A and B, we let $F(A, B) = \sup_{x \in A} d(x, B)$. Then

$$d_H(A, B) = \max\left(F(A, B), F(B, A)\right).$$

We first prove that for any triplet of sets, A, B, C, we have

$$F(A, B) \le F(A, C) + F(C, B). \tag{1.1}$$

The proof is a sequence of applications of the definitions of supremum (least upper bound) and infimum (greatest lower bound). Let $x \in A$ and $z \in C$. By proposition 1.2 we have

$$d(x, B) \le d(x, z) + d(z, B)$$

and therefore

$$d(x, B) \le d(x, z) + F(C, B).$$

Since this holds for all $z \in C$ we have

$$d(x, B) \le d(x, C) + F(C, B),$$

which gives

$$d(x, B) \leq F(A, C) + F(C, B).$$

Again, this holds for all $x \in A$, so that

$$F(A, B) \leq F(A, C) + F(C, B),$$

which is (1.1). From this we obtain

$$F(A, B) \leq d_H(A, C) + d_H(C, B)$$

and, by interchanging A and B, we conclude

$$d_H(A, B) \leq d_H(A, C) + d_H(C, B).$$

$\square$

1.2.5 Cauchy-Schwarz, Hölder, Minkowski

The triangle inequality is usually the only property of a norm that presents any challenge to prove. We shall recount here how it is obtained for the norms $\|x\|_p$ and $\|x\|_\infty$. The reader has probably encountered them in the study of multivariate analysis.

(A) Euclidean length $\|x\|_2$, more simply denoted by $|x|$.

A pleasing geometric proof was given by Euclid (the Elements, I.20). Needless to say we proceed algebraically.

The Euclidean length can be expressed using the inner product (or scalar product) of two coordinate vectors

$$u \cdot v = \sum_{k=1}^{n} u_k v_k, \quad u = (u_1, \ldots, u_n), \quad v = (v_1, \ldots, v_n).$$

We have $|x| = \sqrt{x \cdot x}$ for all $x \in \mathbb{R}^n$. We recall the Cauchy-Schwarz inequality $u \cdot v \leq |u||v|$. We have

$$
\begin{aligned}
|u + v|^2 &= (u + v) \cdot (u + v) \\
&= u \cdot u + 2u \cdot v + v \cdot v \\
&= |u|^2 + 2u \cdot v + |v|^2.
\end{aligned}
$$

Hence

$$|u + v|^2 \leq |u|^2 + 2|u||v| + |v|^2 = \big(|u| + |v|\big)^2,$$

and therefore

$$|u + v| \leq |u| + |v|.$$

(B) The norm $\|x\|_p$.

In this case the triangle inequality is *Minkowski's inequality*. Given that $p \geq 1$ and given real numbers x_i, y_i, $(i = 1, \ldots, n)$, this states that

$$\left(\sum_{i=1}^{n} |x_i + y_i|^p \right)^{\frac{1}{p}} \leq \left(\sum_{i=1}^{n} |x_i|^p \right)^{\frac{1}{p}} + \left(\sum_{i=1}^{n} |y_i|^p \right)^{\frac{1}{p}}.$$

Minkowski's inequality is derived from *Hölder's inequality*, itself a generalisation of the Cauchy-Schwarz inequality. Let x_i, y_i, $(i = 1, \ldots, n)$ be real numbers, and let $p \geq 1, q \geq 1$, with $(1/p) + (1/q) = 1$. Hölder's inequality states that

$$\sum_{i=1}^{n} |x_i y_i| \leq \left(\sum_{i=1}^{n} |x_i|^p \right)^{\frac{1}{p}} \left(\sum_{i=1}^{n} |y_i|^q \right)^{\frac{1}{q}}.$$

Hölder's inequality is usually derived from a basic inequality of analysis, Young's inequality. Let $p \geq 1, q \geq 1$, with $(1/p) + (1/q) = 1$, and let $a > 0$ and $b > 0$. Young's inequality states that

$$ab \leq \frac{a^p}{p} + \frac{b^q}{q}$$

and can be viewed as a general form of the inequality of arithmetic and geometric means.

(C) *Young $\Rightarrow$ Hölder $\Rightarrow$ Minkowski*.

We recall the proof of these implications. Let $x = (x_i)_{i=1}^{n}$, $y = (y_i)_{i=1}^{n}$ be real coordinate vectors and suppose that $x \neq 0$ and $y \neq 0$. By Young's inequality we obtain

$$\sum_{i=1}^{n} \left| \frac{x_i y_i}{\|x\|_p \|y\|_q} \right| \leq \frac{1}{p} \sum_{i=1}^{n} \left| \frac{x_i}{\|x\|_p} \right|^p + \frac{1}{q} \sum_{i=1}^{n} \left| \frac{y_i}{\|y\|_q} \right|^q = \frac{1}{p} + \frac{1}{q} = 1.$$

This gives Hölder's inequality

$$\sum_{i=1}^{n} |x_i y_i| \le \|x\|_p \|y\|_q$$

given that $x \ne 0$ and $y \ne 0$, but it is obvious that the inequality holds if either of these vectors is 0.

The case $p = 1$ of Minkowski's inequality is very simple; so we can assume that $p > 1$. Recall that q satisfies $(1/p) + (1/q) = 1$, which is equivalent to $(p-1)q = p$. Let x and y be real coordinate n-vectors. Using the inequality $|x_i + y_i| \le |x_i| + |y_i|$, together with Hölder's inequality and the formula $(p-1)q = p$, we have:

$$\sum_{i=1}^{n} |x_i + y_i|^p = \sum_{i=1}^{n} |x_i + y_i||x_i + y_i|^{p-1}$$

$$\le \sum_{i=1}^{n} |x_i||x_i + y_i|^{p-1} + \sum_{i=1}^{n} |y_i||x_i + y_i|^{p-1}$$

$$\le \left(\sum_{i=1}^{n} |x_i|^p\right)^{\frac{1}{p}} \left(\sum_{i=1}^{n} |x_i + y_i|^p\right)^{\frac{1}{q}} + \left(\sum_{i=1}^{n} |y_i|^p\right)^{\frac{1}{p}} \left(\sum_{i=1}^{n} |x_i + y_i|^p\right)^{\frac{1}{q}}$$

that is,

$$\sum_{i=1}^{n} |x_i + y_i|^p \le \left(\left(\sum_{i=1}^{n} |x_i|^p\right)^{\frac{1}{p}} + \left(\sum_{i=1}^{n} |y_i|^p\right)^{\frac{1}{p}}\right) \left(\sum_{i=1}^{n} |x_i + y_i|^p\right)^{\frac{1}{q}}$$

If $x + y \ne 0$ then, recalling that $1 - (1/q) = 1/p$, we obtain Minkowski's inequality

$$\left(\sum_{i=1}^{n} |x_i + y_i|^p\right)^{\frac{1}{p}} \le \left(\sum_{i=1}^{n} |x_i|^p\right)^{\frac{1}{p}} + \left(\sum_{i=1}^{n} |y_i|^p\right)^{\frac{1}{p}},$$

but it obviously remains true if $x + y = 0$.

(D) The norm $\|x\|_\infty$.

Let $x = (x_1, \ldots, x_n)$ and $y = (y_1, \ldots, y_n)$ be real coordinate vectors. Then we have

$$|x_i + y_i| \le |x_i| + |y_i| \le \|x\|_\infty + \|y\|_\infty \quad (1 \le i \le n)$$

and so $\|x + y\|_\infty \le \|x\|_\infty + \|y\|_\infty$.

Extending the triangle inequality to complex coordinate vectors is straightforward as Hölder's and Minkowski's inequalities hold for such vectors with no notational change; only the reinterpretation of $|\lambda|$ as modulus of a complex number is needed.

1.2.6   Exercises

1. Check that example (g) defines a metric. It is called the discrete metric.
2. Let $1 \leq p < q$. Show that $\|x\|_\infty \leq \|x\|_q \leq \|x\|_p \leq n^{1/p}\|x\|_\infty$ for all vectors $x \in \mathbb{R}^n$.
3. Calculate the diameters of the following subsets of $\mathbb{R}^n$ with respect to the norm $\|x\|_p$, where $1 \leq p \leq \infty$. The values depend, of course, on p.

   a) The unit cube $[0, 1]^n$.
   b) The unit Euclidean sphere, that is, the set of all coordinate vectors $(x_1, \ldots, x_n)$ that satisfy $\sum_{k=1}^n x_k^2 = 1$.
      Hint. One way is find the maximum of $\sum_{k=1}^n x_k^p$ subject to the constraint $\sum_{k=1}^n x_k^2 = 1$.

4. Let X be the unit Euclidean sphere in the space $\mathbb{R}^3$, that is,

$$X = \left\{(x_1, x_2, x_3) \in \mathbb{R}^3 : x_1^2 + x_2^2 + x_3^2 = 1\right\}.$$

It may be regarded as a subspace of $\mathbb{R}^3$ with the Euclidean metric. There is another way to define a metric in X, relevant to navigation. We let $d(u, v)$ be the length of the shorter great circle arc joining u and v on the unit sphere. Thus $d(u, v) \in [0, \pi]$. Show that d is a metric on X.

Hint. For points u, v and w on the unit sphere let $a = \widehat{vw}$ be the shorter great circle arc joining v and w, and similarly define $b = \widehat{uw}$ and $c = \widehat{uv}$. The *spherical cosine rule* states that

$$\cos a = \cos b \cos c + \cos A \sin b \sin c$$

where A is the angle of the spherical triangle u, v, w at the vertex u. Without loss of generality one may assume that $0 < c \leq b \leq a \leq \pi$.

5. We impose on the integers $\mathbb{Z}$ the dyadic metric $d(x, y) = 2^{-v}$, where 2^v is the highest power of 2 that divides $x - y$, with $d(x, y) = 0$ if $x = y$. This was item (h) in the list of examples of metric spaces. Show that d is a metric on $\mathbb{Z}$ and that it satisfies

$$d(x, y) \leq \max\left(d(x, z), d(y, z)\right).$$

A metric space that satisfies this inequality, which is stronger than the triangle inequality, is called an *ultrametric space*.

6. Ultrametric spaces (see the previous exercise) have counterintuitive properties. Show that every "triangle" is "isosceles". That is, given three points a, b and c, two of the three quantities $d(a, b)$, $d(b, c)$, $d(c, a)$ are equal; more precisely, if they are placed in increasing order then the second and third are equal.

7. Calculate $d_H(A, B)$ (the Hausdorff pseudometric) for the following pairs of subsets of $\mathbb{R}^2$, using the Euclidean metric in item (a) and the metric $\|(x, y)\|_\infty$ in item (b):

a) $A = \{(x, y) : x^2 + y^2 \leq 1\}$,
   $B = \{(x, y) : (x - a)^2 + (y - b)^2 \leq 1\}$

b) $A = \{(x, y) : 0 \leq x \leq 1, \ 0 \leq y \leq 1\}$,
   $B = \{(x, y) : 0 \leq x \leq 2, \ 0 \leq y \leq 2\}$

8. A subset A of a vector space X over $\mathbb{R}$ is said to be convex if it satisfies the following condition: for all x and y in A, and for all real t such that $0 \leq t \leq 1$, the vector $(1 - t)x + ty$ lies in A. In short, A includes the line segment joining any two of its points.

a) Let X be a normed vector space over $\mathbb{R}$. Show that the sets

$$B := \{x \in X : \|x\| < 1\}$$

and

$$\overline{B} := \{x \in X : \|x\| \leq 1\},$$

called the open unit ball and the closed unit ball respectively, are convex.

b) Draw some pictures showing B for $X = \mathbb{R}^2$, with the norms $\|x\|_1$, $\|x\|_2$, $\|x\|_5$ and $\|x\|_\infty$.

9. The connection between convex sets and norms can be explored further. Let X be a vector space over $\mathbb{R}$. Let A be a convex subset of X with the following properties:

i) For each $x \in A$ and for each real t such that $-1 \leq t \leq 1$, the point tx lies in A.

ii) For each $x \neq 0$ there exists $\lambda > 0$ such that $\lambda x \in A$, and $\mu > 0$ such that $\mu x \notin A$.

Define the function $p : X \to [0, \infty[$ by

$$p(x) = \inf\{1/\lambda : \lambda > 0 \wedge \lambda x \in A\}.$$

Prove that $p(x)$ is a norm on X.

Hint. It might help to observe that the set of all $\lambda \geq 0$ such that $\lambda x \in A$ is an interval and infer that it is either $[0, 1/p(x)]$ or $[0, 1/p(x)[$.

10. Let E be a real vector space and let $p : E \to [0, \infty[$ be a function that has all the properties of a norm, except that we do not require that $p(x) = 0$ implies that $x = 0$. Thus, although it is the case that $p(0) = 0$, there may be non-zero vectors x, such that $p(x) = 0$. If so, we call p a *seminorm*. Let $N = p^{-1}(0)$.

   a) Show that N is a linear subspace of E.
   b) Show that a norm can be defined on the quotient space E/N by setting

   $$\|N + x\| = p(x).$$

   One must show that $\|N + x\|$ is well-defined (that $p(x)$ depends only on $N + x$) as well as showing that it is a norm.

1.3   Cantor's Middle Thirds Set

Problems about the convergence of Fourier series and the integrability of functions stimulated an interest in bizarre subsets of the real numbers. One subset in particular of the metric space $\mathbb{R}$ appears very frequently in analysis, as we shall see. Viewed as a metric space in its own right it exhibits some striking properties. In this section we study the Cantor set, the protean nature of which will become apparent later in this text. Along the way we will recall Cantor's famous diagonal argument, which he used to prove the uncountability of the real numbers. This section only uses concepts from fundamental analysis (real analysis) and set theory.

We begin with the real number interval $A_1 = [0, 1]$. From A_1 we remove the open interval (its open middle third) $]\frac{1}{3}, \frac{2}{3}[$. There remains the set

$$A_2 = [0, \tfrac{1}{3}] \cup [\tfrac{2}{3}, 1].$$

From these two intervals we remove their open middle thirds $]\frac{1}{9}, \frac{2}{9}[$ and $]\frac{7}{9}, \frac{8}{9}[$. There remains the set

$$A_3 = [0, \tfrac{1}{9}] \cup [\tfrac{2}{9}, \tfrac{1}{3}] \cup [\tfrac{2}{3}, \tfrac{7}{9}] \cup [\tfrac{8}{9}, 1].$$

We continue in this way, at each step removing the open middle thirds of all the intervals making up A_n, to leave the set A_{n+1}. It is clear that A_n is the union of 2^{n-1} closed intervals, that are pairwise disjoint, and that each of its constituent intervals has length $1/3^{n-1}$. The total length of all the intervals making up A_n is therefore $(2/3)^{n-1}$, which tends to 0 as $n \to \infty$.

The Cantor middle thirds set is defined as

$$B = \bigcap_{n=1}^{\infty} A_n.$$

At first sight it might appear that points in B are rare, it being obvious only that the endpoints of intervals that are removed at each step remain in B. They form a countable set. In fact all such endpoints apart from 0 and 1 are fractions of the form $a/3^m$, but which fractions of this kind are such endpoints is not immediately clear. Another striking fact is that the total length of the intervals removed up to the nth step tends to 1. This is clear because the total length of the intervals making up A_n tends to 0. This also suggests that points in B are rare, and it makes the next proposition all the more surprising.

Proposition 1.3 B *is uncountable.*

Before we prove this proposition we develop a description of the Cantor middle thirds set using ternary fractions. Every point in the interval $[0, 1]$ has a representation as a ternary fraction

$$x = (0.a_1a_2a_3\ldots)_3 = \sum_{k=1}^{\infty} \frac{a_k}{3^k}$$

where each digit is 0, 1 or 2. The ambiguous cases that arise because

$$0.a_1\ldots a_n 0\overline{2} = 0.a_1\ldots a_n 1\overline{0}, \quad 0.a_1\ldots a_n 1\overline{2} = 0.a_1\ldots a_n 2\overline{0}$$

play an essential role in the theory of the Cantor middle thirds set, and we do not wish to fix in advance a preferred representation in these cases. Note that $1 = 0.\overline{2}$. The line over a digit or string of digits, means that the string is to be repeated indefinitely. We omit the subscript '3' when it is obvious we mean a ternary fraction.

The set A_2 is obtained from A_1 by removing all fractions for which $a_1 = 1$ *unavoidably*. Note that $1/3$ is not removed since, although $1/3 = 0.1\overline{0}$, we also have $1/3 = 0.0\overline{2}$, avoiding the digit '1'. Nor is $2/3$ since, though $2/3 = 0.1\overline{2}$, we also have $2/3 = 0.2\overline{0}$, again avoiding the digit '1'. The set A_2 therefore consists of all numbers in $[0, 1]$ that have a ternary representation with first digit 0 or 2, where some stress should fall on the indefinite article.

Next, A_3 consists of numbers that have a ternary representation such that the first digit is 0 or 2 and the second digit is 0 or 2. And so on. We find that A_n consists of numbers that have a ternary representation such that each digit $a_1, \ldots, a_{n-1}$ is either 0 or 2 (even though they may also have a different ternary representation that includes the digit 1).

It should now be clear that the Cantor set B consists of all numbers in the interval $[0, 1]$ that have a ternary representation $0.a_1a_2\ldots$ where each digit is 0 or 2.

Proof of Proposition 1.3. We use Cantor's diagonal argument. If all the elements of B could be listed as a sequence $(t_n)_{n \in \mathbb{N}_+}$ we could obtain a contradiction by constructing a number $t \in B$ that does not appear in the list. Write each number t_n as a ternary fraction using only the digits 0 and 2. We let

$$t = (0.b_1b_2\ldots)_3$$

where $b_n = 0$ if the nth digit of t_n is 2, and $b_n = 2$ if the nth digit of t_n is 0. It is clear that t does not appear in the sequence $(t_n)_{n \in \mathbb{N}_+}$ but t is all the same in B, which is a contradiction.[1]   □

Exercise Show that the intervals that are removed in the construction of Cantor's middle thirds set are of the form

$$](0.a_1 \ldots a_n 0\bar{2})_3, (0.a_1 \ldots a_n 2\bar{0})_3[\; = \;]\frac{(a_1 \ldots a_n 1)_3}{3^{n+1}}, \frac{(a_1 \ldots a_n 2)_3}{3^{n+1}}[$$

where each digit a_k is 0 or 2.

The Cantor set B has an interesting property, that is related to its uncountability, as we shall see later. For each $t \in B$ one may approximate t with elements of B distinct from t, with arbitrarily small error. This property may be expressed in several equivalent ways:

a) For all $t \in B$ and for all $\varepsilon > 0$, there exists $s \in B$, such that $s \neq t$ and $|s - t| < \varepsilon$.
b) For all $t \in B$ and for all $\varepsilon > 0$, the set $B \cap \,]t - \varepsilon, t + \varepsilon[\backslash \{t\}$ is not empty.
c) Every $t \in B$ is a limit point of the set B (limit point is a metric space concept which we define later; the reader may already be familiar with limit points of subsets of $\mathbb{R}$, which is all we use here).

The proof of this property is simple if we use the representation of elements of B as ternary fractions using only the digits 0 and 2. For let $t \in B$ and write

$$t = (0.a_1 a_2 a_3 \ldots)_3$$

where each digit is 0 or 2. Let $\varepsilon > 0$. We construct $s \in B$, such that $s \neq t$ but $|s - t| < \varepsilon$, by choosing a sufficiently large n and changing the nth digit, either from 0 to 2, or from 2 to 0.

The representation of elements of B as ternary fractions, using only the digits 0 and 2, sets up a bijection from B to the set of infinite sequences of the digits 0 and 2. In other words we have a bijection from B to the infinite product $\{0, 2\}^{\mathbb{N}_+}$, which consists of all functions from $\mathbb{N}_+$ to the two element set $\{0, 2\}$. This remains true if '2' is replaced by '1'. We have a bijection from B to the set of all sequences of the digits '0' and '1'. The latter set can be thought of as the set of all subsets of $\mathbb{N}_+$; just think of such a sequence as the characteristic function (or indicator function) of a subset of $\mathbb{N}_+$. The fact that the set of all subsets of $\mathbb{N}_+$ has cardinality **c** is commonly proved in courses on set theory. Therefore, Cantor's middle thirds set has the same cardinality as $\mathbb{R}$.

---

[1] Cantor's argument is not really a case of *reductio ad absurdum* since the assumption, that all elements of B appear in the sequence $(t_n)_{n \in \mathbb{N}_+}$, is never used in the deductive chain. However, it is nearly always presented in this way, and it makes it easy to remember.

Fig. 1.1 The Cantor set as a
binary tree

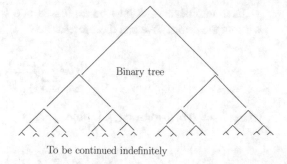

Binary tree

To be continued indefinitely

The inverse of the bijection defined in the last paragraph is easy to describe: it maps the sequence $(a_k)_{k=1}^{\infty}$, where each term is 0 or 1, to the real number

$$\sum_{k=1}^{\infty} \frac{2a_k}{3^k}$$

which clearly lies in B.

Why restrict oneself to numerical digits? We have equally a bijection of B to

$$\{no, yes\}^{\mathbb{N}+}$$

or to

$$\{left, right\}^{\mathbb{N}+}.$$

The last set can be viewed as the set of all paths through an infinite binary tree. It is illustrated in Fig. 1.1. The capacity of the Cantor set to appear in wildly different guises, what we can call its protean quality, is all the more striking because these different manifestations, such as the middle thirds set and the set of paths through an infinite binary tree, not only share the same cardinal, but they share the same topology. What is meant by this will be clarified later in this text.

1.3.1 Exercises

1. Show that Cantor's middle thirds set $B \subset [0, 1]$ does not include any open interval $]c, d[$ (with $c < d$).
2. In this exercise we define a function $f : B \rightarrow [0, 1]$ (where B is Cantor's middle thirds set) that is increasing and surjective. This function is important in the theory of the Lebesgue integral.

Each number $x \in B$ may be expressed in a unique way as a ternary fraction without the digit 1. We can therefore write

$$x = \sum_{n=1}^{\infty} \frac{2b_n}{3^n}$$

where $b_n = 0$ or 1 for each n. Define

$$f(x) = \sum_{n=1}^{\infty} \frac{b_n}{2^n}.$$

a) Show that f is surjective, but not injective.
b) Show that f is increasing, though not strictly increasing. In other words, if $x, y \in B$ and $x < y$, then $f(x) \le f(y)$.
c) Suppose that $x, y \in B$, $f(x) = f(y)$ but $x < y$. Show that x and y are the endpoints of one of the intervals that were removed in the construction of the middle thirds set.
d) Extend f from B to an increasing function on all of $[0, 1]$, by making it constant on each interval that was removed in the construction of the middle thirds set. Show that the extension is continuous.

Hint. The extended function, with domain and codomain $[0, 1]$, is both increasing and surjective. What does analysis say about the discontinuities of an increasing function?

Note. The function f has the counterintuitive property that it has derivative 0 on each of a countable collection of open intervals whose total length is 1. At the same time the graph of f succeeds in climbing up from 0 to 1, taking all values in between.

3. For readers familiar with Lebesgue measure. Show that Cantor's middle thirds set has Lebesgue measure 0.

1.4   The Normed Spaces of Functional Analysis

1.4.1   Sequence Spaces

We extend the norms $\|x\|_p$ to the case when x is a real infinite sequence. The set of all real sequences $a = (a_i)_{i=1}^{\infty}$, that is, the set $\mathbb{R}^{\mathbb{N}+}$, is a vector space over $\mathbb{R}$. Sum and scalar multiplication are defined as for vectors in $\mathbb{R}^n$. Let $a = (a_i)_{i=1}^{\infty}$, $b = (b_i)_{i=1}^{\infty}$, $\lambda \in \mathbb{R}$. We set

$$a + b = (a_i + b_i)_{i=1}^{\infty}, \qquad \lambda a = (\lambda a_i)_{i=1}^{\infty}.$$

In order to define the norms $\| \cdot \|_p$ for infinite sequences we are forced to restrict our attention to certain linear subspaces of $\mathbb{R}^{\mathbb{N}+}$; which subspace will depend on p.

Let $p \geq 1$ be a real number. We define the classical sequence spaces:

a) ℓ^p is the set of all $a \in \mathbb{R}^{\mathbb{N}+}$, such that $\sum_{i=1}^{\infty} |a_i|^p < \infty$.
b) ℓ^{∞} is the set of all bounded sequences $a \in \mathbb{R}^{\mathbb{N}+}$.

For each $a \in \ell^p$ we can define

$$\|a\|_p = \Big(\sum_{i=1}^{\infty} |a_i|^p \Big)^{\frac{1}{p}}.$$

This is a finite number for all $a \in \ell^p$.

Minkowski's inequality $\|a + b\|_p \leq \|a\|_p + \|b\|_p$ is valid for a and b in $\mathbb{R}^{\mathbb{N}+}$, though we must in general allow the value ∞ for the infinite sums. It shows that if a and b are in ℓ^p so is $a + b$. Clearly we also have $\lambda a \in \ell^p$ if $a \in \ell^p$ and $\lambda \in \mathbb{R}$. We see that ℓ^p is a normed vector space carrying the norm $\|a\|_p$.

For all $a \in \ell^{\infty}$ we let

$$\|a\|_{\infty} = \sup_{1 \leq i < \infty} |a_i|.$$

This is a norm on ℓ^{∞} (which is rather obviously a vector space). We use the supremum here, rather than the maximum which we used for $\mathbb{R}^n$, as the set $\{|a_i| : 1 \leq i < \infty\}$, though bounded, does not necessarily have a highest element.

The spaces ℓ^p, $(1 \leq p < \infty)$, and ℓ^{∞} are distinct vector subspaces of the vector space $\mathbb{R}^{\mathbb{N}+}$. Given that $p < q$ they satisfy

$$\ell^p \subsetneqq \ell^q \subsetneqq \ell^{\infty} \subsetneqq \mathbb{R}^{\mathbb{N}+}$$

and, as indicated, all the inclusions are proper. It is left to the reader (see 1.2 exercise 1.2.6) to show that in general, for coordinate vectors $x = (x_1, \ldots, x_n)$, $1 \leq k \leq n$, and $1 \leq p \leq q < \infty$ we have:

$$|x_k| \leq \|x\|_{\infty} \leq \|x\|_q \leq \|x\|_p \leq n^{\frac{1}{p}} \|x\|_{\infty} \tag{1.2}$$

These inequalities remain valid in the limit $n \to \infty$ if we allow ∞ as a possible limit. This implies the inclusions noted above. The reader should show that the inclusions are proper by producing suitable counterexamples to equality.

In the examples just considered we may allow complex numbers. Then we obtain complex vector spaces of sequences in which we may use the norms $\|a\|_p$ and $\|a\|_{\infty}$ defined by the same formulas as before, but interpreting $|a_k|$ as the modulus of a complex number. The symbols ℓ^p can, if thought necessary, be replaced by $\ell^p(\mathbb{R})$ for the real case, and by $\ell^p(\mathbb{C})$ for the complex case. Usually this is not necessary as the context makes it clear which is meant.

The spaces ℓ^p, $(1 \leq p \leq \infty)$, are examples of sequence spaces that are also normed spaces. They become metric spaces when equipped with the metric $d(a, b) = \|a - b\|_p$. But they are proper subsets of $\mathbb{R}^{\mathbb{N}_+}$. We need different methods to define a metric on all of the space $\mathbb{R}^{\mathbb{N}_+}$, as we shall see, beginning with exercise 3 below.

1.4.2   Function Spaces

We can build normed spaces, and therefore metric spaces, out of sets of functions. Let M be a set. The set of all bounded functions $f : M \to \mathbb{R}$ constitutes a vector space $B(M)$ over $\mathbb{R}$. The sum of two functions and scalar multiplication are defined pointwise:

a) $(f + g)(x) = f(x) + g(x)$, $(x \in M)$
b) $(\lambda f)(x) = \lambda f(x)$, $(x \in M)$.

We define a norm on $B(M)$ by letting

$$\|f\|_\infty = \sup_{x \in M} |f(x)|.$$

In the cases that $M = [[1, n]]$ and $M = \mathbb{N}_+$, we obtain the spaces $\mathbb{R}^n$ and ℓ^∞ respectively, with the norm denoted by the common notation $\|a\|_\infty$.

The distance between two bounded functions with domain M is given by

$$d(f, g) = \sup_{x \in M} |f(x) - g(x)|.$$

As we saw, as soon as we have a metric we can express the notions of convergence and limit of a sequence. Let $(f_n)_{n=1}^\infty$ be a sequence in $B(M)$ and let $g \in B(M)$. Then the claim that $\lim_{n \to \infty} f_n = g$ amounts to saying that

$$\lim_{n \to \infty} \sup_{x \in M} |f_n(x) - g(x)| = 0.$$

This is stronger than simply requiring that $\lim_{n \to \infty} f_n(x) = g(x)$ should hold for all $x \in M$. What we want is the following: for all $\varepsilon > 0$ there exists N, such that

$$\sup_{x \in M} |f_n(x) - g(x)| < \varepsilon$$

holds for all $n \geq N$.

The displayed inequality implies that ε is a *strict upper bound for the function* $|f_n - g|$ (which has domain M). The reader has probably encountered the notion of uniform convergence of a sequence of functions having as common domain a subset of $\mathbb{R}$. What we have here is uniform convergence of the sequence $(f_n)_{n=1}^{\infty}$, with common domain M.

Thus the notion of uniform convergence of a sequence of functions can, in the important case when the functions are bounded on their common domain, be expressed as convergence in a certain, rather simple, metric space.

As in the cases of coordinate spaces and sequence spaces there are also complex versions of function spaces. These are built out of functions $f : M \to \mathbb{C}$. We shall not usually adopt any special notation to distinguish the complex versions, allowing the context to indicate whether $B(M)$ is the space of bounded real functions, or the space of bounded complex functions, with domain M.

1.4.3 Spaces of Continuous Functions

In the previous section M was merely a set. Continuity of the functions was not an issue. Now we shall study functions having as domain a set for which continuity of the function can be defined.

Consider first the case when M is the bounded, closed interval $[a, b]$. By the boundedness theorem of fundamental analysis, all continuous functions $f : M \to \mathbb{R}$ are bounded. They form a vector subspace of $B(M)$ which we denote by $C(M)$.

The space $C(M)$ is immensely important in analysis. The norm on $C(M)$ is just the restriction of the norm previously defined for $B(M)$:

$$\|f\|_{\infty} = \sup_{x \in M} |f(x)|.$$

The reader is no doubt aware that continuous functions on the bounded, closed interval $[a, b]$ attain their maximum. Hence the supremum here is actually the maximum value.

We can also consider the space of all continuous functions on an open interval, or we can let M be a subset of $\mathbb{R}$ of a quite arbitrary kind. Continuous functions with domain M are then not necessarily bounded, but we can form the space of all *bounded and continuous* functions with domain M, and make it into a normed space with the norm $\|f\|_{\infty}$. We shall denote this space by $C_B(M)$.

The spaces $C(M)$ and $C_B(M)$ are capable of vast generalisation, in which M is a metric space replacing the real number sets of the previous paragraph. These are studied in chapter 3.

1.4.4 Spaces of Integrable Functions

Functions Riemann integrable on the interval $[a, b]$ form a subspace of the space of bounded functions $B([a, b])$. We shall denote it by $R(a, b)$. We can use the norm

$\|f\|_\infty$ on $R(a, b)$ but it is also possible to use the integral to define a norm. Let $p \geq 1$. We set

$$\|f\|_p = \left(\int_a^b |f|^p \right)^{\frac{1}{p}}, \quad (f \in R(a, b)).$$

The triangle inequality holds here, in the form of *Minkowski's inequality for integrals* (see below). However there is a problem. If $\|f\|_p = 0$ it does not necessarily follow that $f = 0$. The quantity $\|f\|_p$ is only a seminorm (see 1.2 exercise 1.2.6) and the expression $\|f - g\|_p$ only defines a pseudometric (see section 1.2 for the definition).

As with all pseudometrics we can form equivalence classes of the elements, identifying two functions f and g if $\|f - g\|_p = 0$. The equivalence classes then form a normed vector space. Two functions are equivalent if they differ at a finite set of points, but the converse is false. If we restrict our attention to continuous functions only, this problem does not arise, since $\|f\|_p = 0$ then implies $f = 0$. However, it may be inconvenient or unnatural to exclude discontinuous integrable functions. The fact that, when using a space of integrable functions, a "function" is really an equivalence class of functions, causes little inconvenience in practice.

The shortcomings of the space $R(a, b)$ will become apparent. These can be alleviated by using a more advanced integral, the Lebesgue integral.

1.4.5   Hölder's and Minkowski's Inequalities for Integrals

Let f and g be Riemann integrable on $[a, b]$. Then we have:

1) (Hölder) If $p > 1, q > 1$ and $(1/p) + (1/q) = 1$ then

$$\int_a^b |fg| \leq \left(\int_a^b |f|^p \right)^{\frac{1}{p}} \left(\int_a^b |g|^q \right)^{\frac{1}{q}}$$

The case $p = 2$ is the Cauchy-Schwarz inequality for integrals.

2) (Minkowski) If $p \geq 1$ then

$$\left(\int_a^b |f + g|^p \right)^{\frac{1}{p}} \leq \left(\int_a^b |f|^p \right)^{\frac{1}{p}} + \left(\int_a^b |g|^p \right)^{\frac{1}{p}}$$

The proofs of these important inequalities are based on Young's inequality and are left to the reader.

1.4.6 Exercises

1. Extend the inequalities (1.2) to the case of infinitely many coordinates. Use them to prove that if $1 \le p < q < \infty$, then:

$$\ell^p \subset \ell^q \subset \ell^\infty \subset \mathbb{R}^{\mathbb{N}+}.$$

   Show that all the inclusions are proper.

2. a) Let X and Y be normed spaces, with norms $\|x\|_X$ and $\|y\|_Y$, and let $T : X \to Y$ be a linear mapping. For each $x \in X$ define

$$\|x\|' = \|x\|_X + \|Tx\|_Y.$$

   Show that $\|x\|'$ is a norm on X.

   b) An important example of item (a) produces a norm on $C^1[a, b]$. This space consists of all functions f in $C[a, b]$, that are differentiable in the open interval $]a, b[$, such that f' extends to a continuous function in the closed interval $[a, b]$. The C^1-norm is given by

$$\|f\|_{C^1} = \|f\|_\infty + \|f'\|_\infty, \quad (f \in C^1[a, b]).$$

   Describe, in terms of uniform convergence of function sequences, what it means to say that $\lim_{n \to \infty} f_n = g$ in the space $C^1[a, b]$.

3. The set $\mathbb{R}^{\mathbb{N}+}$ is the set of all real sequences $(a_n)_{n=1}^\infty$. It is also a vector space where sequences are added coordinatewise.

   a) Show that one may define a metric on $\mathbb{R}^{\mathbb{N}+}$ by setting

$$D(a, b) = \sum_{n=1}^\infty 2^{-n} \left(\frac{|a_n - b_n|}{1 + |a_n - b_n|} \right),$$

   where $a = (a_n)_{n=1}^\infty$ and $b = (b_n)_{n=1}^\infty$.

   b) Show that $D(a, b) = D(a - b, 0)$, but there is no norm such that $D(a, b) = \|a - b\|$.

   c) Other metrics, of which item (a) is a special case, may be defined on $\mathbb{R}^{\mathbb{N}+}$. We fix positive numbers β_n such that $\sum_{n=1}^\infty \beta_n < \infty$, and a function $f : [0, \infty[\to [0, \infty[$ with the properties described in 1.1 exercise 5, with the additional properties that f is bounded above and continuous at 0. We then set

$$D'(a, b) = \sum_{n=1}^\infty \beta_n f(|a_n - b_n|), \quad (a = (a_n)_{n=1}^\infty, \ b = (b_n)_{n=1}^\infty)$$

   Verify that that D' defines a metric on $\mathbb{R}^{\mathbb{N}+}$.

Note. A common example differing from that of item (a) is formed with $f(x) = \min(x, 1)$. The reason for requiring continuity at 0 is connected with the desired properties of convergence in $\mathbb{R}^{\mathbb{N}+}$ and will be apparent later (section 3.2).

4. Prove Hölder's inequality for integrals.

   Hint. Let $A = \left(\int_a^b |f|^p \right)^{1/p}$ and $B = \left(\int_a^b |g|^q \right)^{1/q}$ and apply Young's inequality $st \leq s^p/p + t^q/q$ with $s = |f(x)|/A$ and $t = |g(x)|/B$.

5. Prove Minkowski's inequality for integrals. You may assume that the integrals are Riemann integrals.

6. a) Let $1 < p < \infty$. Prove the inequalities:

$$|x + y|^p + |x - y|^p \leq 2^{p-1}\left(|x|^p + |y|^p\right), \quad (2 \leq p < \infty)$$

$$|x + y|^p + |x - y|^p \leq 2\left(|x|^p + |y|^p\right), \quad (1 < p < 2)$$

   for any pair of real numbers x and y.

   Hint. For $2 \leq p < \infty$ the function $f(t) = |t|^{p/2}$ is convex and satisfies $f(0) = 0$. Recall some properties of such functions. See also 1.1 exercise 5. For $1 < p < 2$ the function $f(t) = |t|^{p/2}$ is concave and the reverse of the first inequality is obtained, from which the second results on replacing x by $x + y$ and y by $x - y$.

   b) Let $1 < p < \infty$ and let q satisfy $(1/p) + (1/q) = 1$. Prove that for vectors x and y in $\mathbb{R}^n$ or $\mathbb{R}^{\mathbb{N}+}$ we have

$$\|x + y\|_p^p + \|x - y\|_p^p \leq 2^{p-1}\left(\|x\|_p^p + \|y\|_p^p\right), \quad (2 \leq p < \infty)$$

$$\|x + y\|_p^p + \|x - y\|_p^p \leq 2\left(\|x\|_p^p + \|y\|_p^p\right), \quad (1 < p < 2)$$

$$\|x + y\|_p^q + \|x - y\|_p^q \leq 2\left(\|x\|_p^p + \|y\|_p^p\right)^{q/p}, \quad (1 < p < 2)$$

   These results are called *Clarkson's inequalities* and play an important role in the geometry of normed spaces. See the next item.

   Hint. The first and second follow directly from item (a). The third follows from the second using the fact that the function $f(t) = |t|^{q/p}$ is convex in the case $1 < p < 2$ and satisfies $f(0) = 0$.

   c) Let $1 < p < \infty$, let X be $\mathbb{R}^n$ with the norm $\|x\|_p$, or ℓ^p. Show that for all ε in the range $0 < \varepsilon < 2$, there exists $\delta > 0$, such that if x and y are vectors such that $\|x\| = \|y\| = 1$ and $\|x - y\| > \varepsilon$, then

$$\left\|\tfrac{1}{2}(x + y)\right\| < 1 - \delta.$$

Note. This property of the normed space X is called uniform convexity. It is true more widely for all L^p-spaces (a topic of integration theory) with $1 < p < \infty$; the spaces X of this exercise are simple cases. All Hilbert spaces (an important type of normed space where one can do Euclidean geometry) are uniformly convex.

Chapter 2
Basic Theory of Metric Spaces

> *I was at the mathematical school, where the master taught his pupils by a method scarce imaginable to us in Europe. The proposition and demonstration were fairly written on a thin wafer, with ink composed of a cephalic tincture. This the student was to swallow upon a fasting stomach, and for three days following eat nothing but bread and water. As the wafer digested, the tincture mounted to his brain bearing the proposition with it.*
>
> J. Swift. *Gulliver's Travels: A Voyage to Laputa.*

In the previous chapter we saw that metric spaces are very common in analysis. In this chapter we develop the basic machinery of metric spaces. We shall encounter many ideas that are familiar from fundamental analysis, but in a more general and abstract context. Examples are open and closed sets, convergence of sequences, continuous functions and completeness.

2.1 Balls in a Metric Space

Let X be a set equipped with a metric d. We define open balls and closed balls in X. Let $a \in X$ and $r > 0$:

a) The *open ball* with centre a and radius r is the set

$$B_r(a) = \{x \in X : d(a, x) < r\}.$$

b) The *closed ball* with centre a and radius r is the set

$$B_r^-(a) = \{x \in X : d(a, x) \leq r\}.$$

R. Magnus, *Metric Spaces*, Springer Undergraduate Mathematics Series,
https://doi.org/10.1007/978-3-030-94946-4_2

Obviously

$$B_s(a) \subset B_s^-(a) \subset B_r(a) \subset B_r^-(a)$$

whenever $0 < s < r$.

Various notations for balls are used in the literature, mostly using an upper case 'B'. Thus the open ball has been denoted by $B(a, r)$, or sometimes $S(a, r)$ ('S' probably for 'sphere'). There is little agreement over notation for the closed ball. The superscript '$-$' is supposed to remind you of a saucepan lid.

The correct syntax when speaking is to say "$B_r(a)$ is a ball in the metric space (X, d)." We might shorten this to "$B_r(a)$ is a ball in the metric space X", or even "$B_r(a)$ is a ball in X" (as we did above), but only if the context makes it clear that X is equipped with a metric, which, for brevity, we do not mention. It is never correct to say "$B_r(a)$ is a ball in the set X", unless it should happen that $B_r(a)$ is an element of X.

2.1.1 Limit of a Convergent Sequence

Let $(x_n)_{n=1}^{\infty}$ be a sequence of points in the metric space (X, d), (that is, a sequence in X, the latter being equipped with a metric d). The definition of limit has already been given. A recurrent theme of metric space theory is to rewrite the same concept in different ways. Now we give the definition again, but express it in terms of balls.

Definition A point $a \in X$ is said to be the limit of the sequence $(x_n)_{n=1}^{\infty}$ if the following condition is satisfied: for all $\varepsilon > 0$ there exists $N \in \mathbb{N}_+$, such that $x_n \in B_\varepsilon(a)$ for all $n \geq N$.

We write

$$\lim_{n \to \infty} x_n = a,$$

or more informally $x_n \to a$, to express this. We say that the sequence $(x_n)_{n=1}^{\infty}$ is convergent in X if it has a limit. The use, in the definition of limit, of the definite article 'the' will be justified shortly.

The definition of limit can be expressed in our simplified first order logic:

$$(\forall \varepsilon \in \mathbb{R}_+)(\exists N \in \mathbb{N})(\forall n \in \mathbb{N})\big(n \geq N \Rightarrow x_n \in B_\varepsilon(a)\big)$$

Note that $x_n \in B_\varepsilon(a)$ is equivalent to $d(x_n, a) < \varepsilon$. The latter replaces and generalises the inequality $|x_n - a| < \varepsilon$ of fundamental analysis, used in the definition of limit of a real number sequence, in which x_n and a are, of course, real numbers.

To verify that $\lim_{n \to \infty} x_n = a$ we have to show that a suitable N can be produced for each positive ε. But actually we don't have to consider all possible ε; a certain countable set of them will do.

Proposition 2.1 *The limit of the sequence $(x_n)_{n=1}^{\infty}$ equals a if and only if the following condition is satisfied: for all $m \in \mathbb{N}_+$ there exists $N \in \mathbb{N}_+$, such that $x_n \in B_{1/m}(a)$ for all $n \geq N$.*

Proof The condition is obviously necessary, To show that it is sufficient we let $\varepsilon > 0$. Let m be an integer such that $m > 1/\varepsilon$. Then $B_{1/m}(a) \subset B_{\varepsilon}(a)$. Let N be an integer, such that $x_n \in B_{1/m}(a)$ for all $n \geq N$. For such n we also have $x_n \in B_{\varepsilon}(a)$. □

2.1.2 Uniqueness of the Limit

We need to show that if $(x_n)_{n=1}^{\infty}$ is convergent then it has a unique limit. This was implicit in the use of the definite article in defining *the* limit. It is required to show that if $\lim_{n \to \infty} x_n = a$, and also $\lim_{n \to \infty} x_n = b$, then $a = b$. We use the following important proposition, the proof of which is the first indication of why we need the triangle inequality.

Proposition 2.2 *Let a and b be points in X, such that $a \neq b$. Then there exists $r > 0$, such that*

$$B_r(a) \cap B_r(b) = \emptyset.$$

Proof By property 1 of metrics, $d(a, b) \neq 0$. Hence there exists r, such that $0 < r < \frac{1}{2}d(a, b)$. If $x \in B_r(a) \cap B_r(b)$, then $d(x, a) < r$ and $d(x, b) < r$, and so, by the triangle inequality,

$$d(a, b) \leq d(x, a) + d(x, b) < 2r < d(a, b),$$

which is impossible. We conclude that $B_r(a) \cap B_r(b)$ is empty. □

Consider now a sequence $(x_n)_{n=1}^{\infty}$, suppose that $a \neq b$, and let r be positive and satisfy $B_r(a) \cap B_r(b) = \emptyset$. It is clearly impossible that any N should exist, such that for all $n \geq N$ we have both that $x_n \in B_r(a)$ and that $x_n \in B_r(b)$. For the two balls have no common point. It follows therefore that it is impossible to have both $\lim_{n \to \infty} x_n = a$ and $\lim_{n \to \infty} x_n = b$.

2.1.3   Neighbourhoods

Let (X, d) be a metric space.

Definition Let $a \in X$. By a *neighbourhood of a* we mean a set $U \subset X$, such that there exists $r > 0$, such that $B_r(a) \subset U$.

Some sources require a neighbourhood to be an open set (a concept defined shortly in section 2.2) and this can cause some confusion if one dips into the middle of a book on topology. The definition here is definitely the most common. It obviously implies that if U is a neighbourhood of a, then any set V, such that $U \subset V$, is also a neighbourhood of a.

Limits can also be understood by using neighbourhoods. The reader can supply the proof of the following proposition.

Proposition 2.3 *The sequence $(x_n)_{n=1}^{\infty}$ is convergent with limit a if and only if the following condition is satisfied: for every neighbourhood U of the point a there exists N, such that $x_n \in U$ for all $n \geq N$.*

2.1.4   Bounded Sets

Let (X, d) be a metric space. The following concept was introduced in 1.1 exercise 3. We restate it, in terms of balls:

Definition A set $A \subset X$ is said to be bounded if there exist $a \in X$ and $r > 0$, such that $A \subset B_r(a)$. A family $(x_i)_{i \in I}$ is said to be bounded if the set $\{x_i : i \in I\}$ is bounded.

If M is bounded then the set of numbers $\{d(x, y) : x \in M, y \in M\}$ is bounded, or equivalently the diameter of M is finite. The converse is also true. The proofs of these assertions, which depend strongly on the triangle inequality, are left to the exercises (again, see 1.1 exercise 3).

Proposition 2.4 *Every convergent sequence is bounded.*

Proof Suppose that $\lim_{n \to \infty} x_n = a$. There exists N, such that $x_n \in B_1(a)$ for all $n \geq N$. Letting r be greater than 1 and greater than each of the finite set of numbers $d(x_i, a), i = 1, \ldots N - 1$, we see that $x_n \in B_r(a)$ for all n.   □

2.1.5   Completeness; a Key Concept

Let (X, d) be a metric space.

Proposition 2.5 *Every convergent sequence* $(x_n)_{n=1}^{\infty}$ *in X has the following property: for all* $\varepsilon > 0$ *there exists N, such that* $d(x_n, x_m) < \varepsilon$ *for all n and m that satisfy* $n \geq N$ *and* $m \geq N$.

Proof Suppose that $\lim_{n \to \infty} x_n = a$. Let $\varepsilon > 0$. Choose N, such that $x_n \in B_{\varepsilon/2}(a)$ for all $n \geq N$. If $n \geq N$ and $m \geq N$ we have, by the triangle inequality,

$$d(x_n, x_m) \leq d(x_n, a) + d(x_m, a) < \frac{\varepsilon}{2} + \frac{\varepsilon}{2} = \varepsilon.$$

$\square$

A sequence that has the property described in proposition 2.5 (irrespective of whether it is convergent or not) is called a Cauchy sequence. We highlight the notion of Cauchy sequence because of its great importance.

Definition A sequence $(x_n)_{n=1}^{\infty}$ in X is called a Cauchy sequence if it has the following property:

For all $\varepsilon > 0$ *there exists N, such that* $d(x_n, x_m) < \varepsilon$ *for all n and m that satisfy* $n \geq N$ *and* $m \geq N$.

In a Cauchy sequence the terms are getting close to each other in a precise sense. When we say that $d(x_n, x_m) < \varepsilon$ for all n and m that satisfy $n \geq N$ and $m \geq N$, we impose no limit on how big $|m - n|$ can become. This can be reflected in an equivalent version of the definition that throws into relief the boundlessness of $|m - n|$. The sequence $(x_n)_{n=1}^{\infty}$ is a Cauchy sequence if it satisfies the following condition:

For all $\varepsilon > 0$ *there exists N, such that* $d(x_{n+p}, x_n) < \varepsilon$ *for all* $n \geq N$ *and all natural numbers p.*

Proposition 2.5 says that all convergent sequences are Cauchy sequences. A given metric space (X, d) does not necessarily have the property that all its Cauchy sequences are convergent. The reader probably knows that for the space of real numbers $\mathbb{R}$, with its usual metric $d(s, t) = |s - t|$, the property of being a Cauchy sequence is called Cauchy's condition, and that all Cauchy sequences in $\mathbb{R}$ are convergent. The latter is Cauchy's convergence principle,[1] and it enables one to see that a sequence of real numbers is convergent without having a candidate for its limit. We have to verify that its terms are getting closer together in the right sort of way, but the verification does not require looking outside the sequence.

We can consider the set of rationals $\mathbb{Q}$ as a subspace of $\mathbb{R}$, with the same metric but restricted to pairs of rationals. A sequence of rational approximations to $\sqrt{2}$ is a Cauchy sequence in $\mathbb{Q}$, because it is a Cauchy sequence in $\mathbb{R}$, but in the metric space $\mathbb{Q}$ it is not convergent because $\sqrt{2}$ is not rational. Thus it appears that the

---

[1] Often called Cauchy's criterion. However, it is unclear whether "Cauchy's criterion" refers to the principle or the condition. It is an advantage to distinguish the two.

metric space $(\mathbb{Q}, d)$, where $d(s, t) = |s - t|$, has a Cauchy sequence that is not convergent.

We next frame an absolutely key property that metric spaces may or may not possess.

Definition A metric space (X, d) is said to be *complete* if all its Cauchy sequences are convergent.

The reader no doubt knows that the real numbers constitute a *complete ordered field*. This is a different notion of completeness. It can be expressed by saying that non-empty sets, that are bounded above, have a least upper bound (and in many other, but equivalent, ways), and it is applicable to *ordered fields*. It is more precise to call it *Dedekind completeness*. Now we are are concerned with a different notion—we can call it *metric completeness* if emphasis is needed—and it is applicable to metric spaces. The realms of metric spaces and ordered fields are distinct, so the two types of completeness are distinct. To be completely explicit we have two concepts:

1. For metric spaces: metric completeness. Every Cauchy sequence is convergent.
2. For ordered fields: Dedekind completeness. Every non-empty set bounded above has a least upper bound.

Cauchy's convergence principle in fundamental analysis asserts that the *metric space* $\mathbb{R}$ is metrically complete, and this is a consequence, usually proved in first courses in analysis, of the fact that the *ordered field* $\mathbb{R}$ is Dedekind complete.

To complicate matters, Cauchy's condition in fundamental analysis makes sense, not just in $\mathbb{R}$, but in any ordered field, because, firstly, the notions of positivity and absolute value are available, the latter defined by $|a| = \max(a, -a)$; and secondly, convergence can be expressed through the ordering (see section 2.6.1 for more on this). There is therefore a third concept, which does not seem to have a name:

3. For ordered fields. Every sequence that satisfies Cauchy's condition is convergent.

This condition is strictly weaker than Dedekind completeness, a fact that sometimes leads to misunderstandings.

Most important metric spaces are complete. Examples of complete metric spaces (references are to the definitions in chapter 1) are:

a) $\mathbb{R}^n$, $n \in \mathbb{N}_+$, with any of the metrics $\|x - y\|_p$ ($1 \le p \le \infty$), (section 1.2).
b) ℓ^p, with $p \ge 1$ or $p = \infty$ (section 1.4).
c) $C[a, b]$ with the metric $\|f - g\|_\infty$ (section 1.4).
d) Cantor's middle thirds set B (section 1.3).

Proofs that these spaces are complete will be given in the next chapter. In contrast, the space $R[a, b]$ with the metric $d(f, g) = \int_a^b |f - g|$ (section 1.4) is not complete. This is a shortcoming of the Riemann integral, rectified by passing to the Lebesgue integral.

2.1.6   Exercises

1. Let (X, d) be a metric space and fix a point p in X. Show that a subset A is bounded if and only if there exists $r > 0$, such that $A \subset B_r(p)$.

2. Identify the balls $B_r(a)$ and $B_r^-(a)$ for the following spaces (X, d), and the given a and r:

   a) $X = \mathbb{N}$ with $d(x, y) = |x - y|, a = 1, r = 1$.
   b) $X = \mathbb{N}$ with $d(x, y) = \min(|x - y|, 1), a = 1, r = 1$.
   c) $X = \mathbb{Z}$ with the dyadic metric (defined in example (h), section 1.2), $a = 0$, $r = \frac{1}{2}$.

3. If X is a vector space and $A \subset X$ then the set $A + x$ is defined by

$$A + x = \{a + x : a \in A\}.$$

   a) Let X be a normed vector space, let $x \in X$ and $r > 0$. Show that

$$B_r(x) = B_r(0) + x.$$

   b) Let $X = \mathbb{R}^{\mathbb{N}+}$ with any of the metrics described in 1.4 exercise 3. Show that the same property holds as in item (a).

   Note. The property that is asked about in this exercise will hold in any vector space with a metric that satisfies $d(x, y) = d(x - y, 0)$. The metric is said to be translation invariant.

4. Find the limits of the following sequences $(x_n)_{n=1}^{\infty}$ in the given metric spaces (X, d), if the limit exists:

   a) $X = \{1/n : n \in \mathbb{N}_+\}, d(x, y) = |x - y|, x_n = 1/n$.
   b) $X = \{0\} \cup \{1/n : n \in \mathbb{N}_+\}, d(x, y) = |x - y|, x_n = 1/n$.
   c) $X = \mathbb{Z}, d(x, y) = \min(|x - y|, 1), x_n = 2^n$.
   d) $X = \mathbb{Z}$ with the dyadic metric (1.2 example (h)), $x_n = 2^n$.

5. Show that a Cauchy sequence is always bounded.

6. Show that a Cauchy sequence that has a convergent subsequence is itself convergent.

7. Equip X with the *discrete metric*: $d(x, y) = 1$ if $x \neq y, d(x, x) = 0$. Show that a sequence $(x_n)_{n=1}^{\infty}$ in X is convergent if and only if it is eventually constant, that is, if and only if there exists N, such that $x_n = x_N$ for all $n \geq N$.

8. Let a and b be points in a metric space, and let $r > 0$ and $s > 0$. Assume that $a \notin B_s(b)$ and $b \notin B_r(a)$. Prove that $B_{r/2}(a) \cap B_{s/2}(b) = \emptyset$.

9. In this exercise and the succeeding one we explore properties of ultrametric spaces (see 1.2 exercises 1.2.6 and 1.2.6). These seem counterintuitive, forcing one to leave the comfort zone of normed vector spaces. Given that (X, d) is an ultrametric space prove the following:

a) If the open balls $B_r(a)$ and $B_r(b)$ have a common point then $B_r(a) = B_r(b)$.

b) If $x \in B_r(a)$ then $B_r(x) = B_r(a)$. Putting it differently: every point in the ball $B_r(a)$ can be viewed as its centre.

c) If $s < r$ and the open balls $B_r(a)$ and $B_s(b)$ have a common point then $B_s(a) \subset B_r(b)$.

10. A primary example of an ultrametric space is the integers $\mathbb{Z}$ with the dyadic metric. This can be extended to the rationals $\mathbb{Q}$ in the following way. We set $d(x, y) = 0$ if $x = y$ and $d(x, y) = 1$ if $|x - y| = 1$. In all other cases we write $|x - y| = 2^u(a/b)$ where u is an integer and a and b are odd integers. We then set $d(x, y) = 2^{-u}$.

   a) Show that this defines a metric that makes $\mathbb{Q}$ into an ultrametric space.

   b) Let $\mathbb{Q}$ be equipped with the dyadic metric. Calculate the sums of the following infinite series, using the usual definition as the limit of the sequence of partial sums, when it exists, but calculating the limit with respect to the dyadic metric:

$$\sum_{k=0}^{\infty} 2^k, \quad \sum_{k=1}^{\infty} 2^k, \quad \sum_{k=0}^{\infty} 4^k.$$

Note. These calculations should dispel habits of thought acquired from doing traditional analysis. The number 2 can be replaced by any prime p. There results the system of p-adic rationals. These form the basis of a highly interesting but somewhat esoteric alternative to the real number system. Every p-adic rational can be expanded in a series

$$\sum_{k=N}^{\infty} a_k p^k,$$

convergent in the p-adic metric. Here the integer N can be negative, while the coefficients a_k are integers in the range 0 to $p - 1$. Not all series of this form converge to rationals. Those that don't converge to p-adic irrationals, the set of which, together with the p-adic rationals, constitutes a field built by a process of completion from the rationals, analogous to the manner in which the normal reals are built out of the rationals.

11. Prove the Vitali covering lemma (finite case). Let $\left(B_{r_k}(a_k)\right)_{k=1}^{m}$ be a finite sequence of balls in a metric space X, such that their union is all of X (in short, the balls *cover* X). Show that there exists a set $S \subset [[1, m]]$ of integers, such that the balls

$$B_{r_k}(a_k), \quad (k \in S)$$

are pairwise disjoint, and the balls

$$B_{3r_k}(a_k), \quad (k \in S)$$

cover X.

Hint. Produce S as a sequence $(j_k)_{k=1}^{\ell}$ in $[[1, m]]$ defined by induction. Here is a verbal description. First choose a ball with the biggest radius. At each step choose a ball with the biggest radius that does not meet any of the balls already chosen. Of course, the challenge is to express this precisely and show that it works.

12. Prove the Vitali covering lemma (infinite case). Let X be a metric space, let $A \subset X$ and let $(r_a)_{a \in A}$ be a *bounded* family of positive real numbers, indexed by A, such that the family of balls

$$\left(B_{r_a}(a)\right)_{a \in A}$$

covers X. Show that there exists a subset $S \subset A$, such that the balls

$$B_{r_a}(a), \quad (a \in S)$$

are pairwise disjoint, and the balls

$$B_{5r_a}(a), \quad (a \in S)$$

cover X.

Hint. Construct a sequence of subsets $S_n \subset A$, $(n \in \mathbb{N}_+)$, using the following (loosely described) inductive procedure. Let $M = \sup_{a \in A} r_a$. Using Zorn's lemma one finds a maximal pairwise disjoint subfamily $(B_{r_a}(a))_{a \in S_1}$ of $(B_{r_a}(a))_{a \in A}$ for which the radii satisfy $2^{-1}M < r_a \leq M$. At the nth step one finds a maximal pairwise disjoint subfamily $(B_{r_a}(a))_{a \in S_n}$ of $(B_{r_a}(a))_{a \in A}$ comprising balls that do not meet any of the balls already chosen, and for which the radii satisfy $2^{-n}M < r_a \leq 2^{-n+1}M$. The set $S := \bigcup_{n=1}^{\infty} S_n$ has the required properties.

2.2   Open Sets, and Closed

The principal use of balls, and by extension neighbourhoods, in a metric space, beyond that of defining convergence, is to define the concepts of open set and closed set.

2.2.1 Open Sets

Let (X, d) be a metric space.

Definition A set $A \subset X$ is said to be open if it is a neighbourhood of each of its points.

An equivalent formulation that merely inserts the definition of neighbourhood, and is what is often used in proofs, is this:

A set $A \subset X$ is open if and only if, for each $a \in A$, there exists $r > 0$, such that $B_r(a) \subset A$.

In the metric space (X, d) the set X is a neighbourhood of all points $x \in X$; it follows that X is open. For a set not to be open it must contain a point x of which it is not a neighbourhood. In particular it must contain a point. It follows that the empty set $\emptyset$ is open.

In the space $\mathbb{R}$ with its usual metric the interval $]a, b[$ (called an open interval, fortunately) is open, but the sets $[a, b]$ and $[a, b[$ are not open.

The property of a set A of being open supposes that A is viewed as a subset of a given set X and that X is assigned a metric d. The most correct syntax is to say "A is an open set *in the metric space* (X, d)". We may sometimes say simply "A is an open set in X", or even "A is open in X" when it is understood that a metric space (X, d) is meant and the metric d is unambiguously known to the reader (or listener). If the metric is not clear this can be considered a little sloppy. When we make a statement about a set A, such as that A is open, we may sometimes have to specify the metric space in which we are viewing A, if a choice of different spaces is possible or if it is not clear from the context.

Proposition 2.6 *An open ball $B_r(a)$ is an open set.*

Proof This is another example showing the importance of the triangle inequality. Let $b \in B_r(a)$. Choose s, such that $0 < s < r - d(a, b)$. Then $B_s(b) \subset B_r(a)$, because if $x \in B_s(b)$ we have

$$d(x, a) \leq d(x, b) + d(b, a) < s + d(b, a) < r.$$

$\square$

A picture of the proof (Fig. 2.1), illustrating $B_r(a)$ and $B_s(b)$ as discs in a Euclidean plane, would enable us to guess that the condition $0 < s < r - d(a, b)$

Fig. 2.1 Guessing the proof from a picture

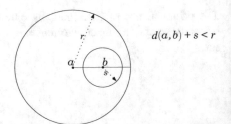

$d(a, b) + s < r$

leads to $B_s(b) \subset B_r(a)$. This can often be a ready means to guess a proof, when the actual verification uses the triangle inequality. This intuition can of course be unreliable, if it uses something more essential to Euclidean geometry, such as angles.

2.2.2 Union and Intersection of Open Sets

Let (X, d) be a metric space.

Proposition 2.7 *Let $(A_i)_{i \in I}$ be a family of subsets of X, such that each set A_i is open. Then $W := \bigcup_{i \in I} A_i$ is an open set.*

Proof Let $x \in W$. Then there exists i, such that $x \in A_i$. But A_i is open and $A_i \subset W$, so that W is a neighbourhood of x. □

The index set I can be infinite, even uncountable. There is no restriction on the cardinality of the family. When we turn to intersection we see a different story.

Proposition 2.8 *Let $(A_i)_{i=1}^n$ be a finite family of open sets in X. Then $W := \bigcap_{i=1}^n A_i$ is open.*

Proof Let $x \in W$. Then $x \in A_i$ for $i = 1, 2, \ldots, n$. For each $i \in \{1, \ldots, n\}$ there exists $r_i > 0$, such that $B_{r_i}(x) \subset A_i$. Let $r = \min\{r_1, \ldots, r_n\}$. Then $r > 0$ and $B_r(x) \subset A_i$ for each i, so that $B_r(x) \subset W$. □

The key point in the proof is the distinction between a finite family of positive numbers and an infinite one; the finite family has a positive lower bound. Not so, in general, for the infinite one. It is left to the reader to find an example showing that the intersection of an infinite family of open sets need not be open.

2.2.3 Closed Sets

Let (X, d) be a metric space.

Definition A set $M \subset X$ is said to be closed if its complement $M^c := X \setminus M$ is open.

Two equivalent formulations, that are often used in arguments about closed sets, are as follows:

A set M is closed if and only if M^c is a neighbourhood of each of its points.

Or, including the definition of neighbourhood:

For each $a \notin M$ there exists $r > 0$, such that $M \cap B_r(a) = \emptyset$.

We see at once that $\emptyset$ and X are both closed, as well as being both open. The fact that a set can be both open and closed can give rise to some merriment. Later we shall specifically study spaces in which there exist sets, that are both open and closed, in addition to $\emptyset$ and X. Therefore we need a nice way to use 'open and closed', as an adjective attributively, as in 'an open and closed set'. This needs to be avoided as 'open and closed sets' would usually mean the same as 'open sets and closed sets', which is not intended here. Many sources use the formation 'clopen' for 'open and closed'. This is unambiguous but hideous. In this text we shall adopt instead the formation 'open-closed set', as in '$\emptyset$ is an open-closed set.'

As was the case for the notion of open set, when we say that a set is closed we are presupposing that it is viewed as a subset of a set X and that X has been assigned a metric. Pedantically correct is to say "M is a closed set in the metric space (X, d)".

Proposition 2.9 *A closed ball $B_r^-(a)$ is a closed set.*

Proof Let $b \notin B_r^-(a)$. We use the second reformulation of the definition of closed, and seek $s > 0$, such that $B_r^-(a) \cap B_s(b) = \emptyset$. This is another proof that can be "discovered" by drawing a picture of Euclidean discs to find the desired s. We have $d(b, a) > r$ and choose s, such that $0 < s < d(b, a) - r$. Now we must have $B_r^-(a) \cap B_s(b) = \emptyset$, for if $x \in B_r^-(a) \cap B_s(b)$ we would find

$$d(a, b) \leq d(a, x) + d(x, b) < r + s < d(a, b),$$

which is impossible.   □

2.2.4   Union and Intersection of Closed Sets

The intersection of a family of closed sets is closed, and no cap is needed on the cardinality of the family. The union of a finite family of closed sets is closed. The union of an infinite family of closed sets is not necessarily closed. Here are the formal statements of these important rules. They follow from the definition of closed set, de Morgan's rules and the corresponding rules for open sets. We assume a metric space (X, d).

Proposition 2.10 *Let $(A_i)_{i \in I}$ be a family of sets such that each set A_i is closed in X. Then $W := \bigcap_{i \in I} A_i$ is closed.*

Proof We have $W^c := \bigcup_{i \in I} A_i^c$, which is open.   □

Proposition 2.11 *Let $(A_i)_{i=1}^n$ be a finite family of sets each of which is closed in X. Then $W := \bigcup_{i=1}^n A_i$ is closed.*

Proof We have $W^c = \bigcap_{i=1}^n A_i^c$, which is open.   □

2.2.5 Characterisation of Open and Closed Sets by Sequences

One of the features of metric space theory is the sometimes heavy use made of sequences. This carries over into analysis, and most particularly, functional analysis.

Let $(a_n)_{n=1}^{\infty}$ be a sequence in a set X, and let $W \subset X$. We say that the sequence $(a_n)_{n=1}^{\infty}$ is eventually in W if there exists N, such that $a_n \in W$ for all $n \geq N$. This is the same as saying that all the terms of the sequence, except at most a finite number of them, are in W. If it is not the case that the sequence $(a_n)_{n=1}^{\infty}$ is eventually in W, (a situation that would naturally be expressed by the ambiguous phrase '$(a_n)_{n=1}^{\infty}$ is not eventually in W'), then there exists a subsequence $(a_{k_n})_{k=1}^{\infty}$ that is wholly outside W. It is left to the reader to construct such a subsequence using the well-ordering of the integers; the axiom of choice is not (yet) needed.

Now we suppose (X, d) to be a metric space. The claim made of a sequence $(a_n)_{n=1}^{\infty}$, that $\lim_{n \to \infty} a_n = b$, can be reduced to the following brief condition:

> For every neighbourhood U of b the sequence $(a_n)_{n=1}^{\infty}$ is eventually in U.

We can use this to characterise open sets by means of sequences.

Proposition 2.12 *Let $W \subset X$. The following are equivalent:*

1) *W is open.*
2) *If the sequence $(a_n)_{n=1}^{\infty}$ in X is convergent and $\lim_{n \to \infty} a_n \in W$, then $(a_n)_{n=1}^{\infty}$ is eventually in W.*

Proof (1) $\Rightarrow$ (2). Suppose that W is open and let $(a_n)_{n=1}^{\infty}$ be a sequence that is convergent and whose limit b is in W. Then W is a neighbourhood of b and so $(a_n)_{n=1}^{\infty}$ is eventually in W.

$\neg$(1) $\Rightarrow$ $\neg$(2). Suppose that W is not open. Then there exists $b \in W$, such that W is not a neighbourhood of b. It follows that for all $r > 0$, we have $B_r(b) \not\subset W$ and so $B_r(b)$ must contain a point not in W. In particular for all $n \in \mathbb{N}_+$ there exists a point a_n in $B_{1/n}(b)$ that is not in W. But then $\lim_{n \to \infty} a_n = b$ but $(a_n)_{n=1}^{\infty}$ is wholly outside W. Note that the axiom of choice was used[2] (where?). $\square$

In the statement of item 2 of the proposition, an intended quantification over all sequences is left unstated. The prelude 'For all sequences $(a_n)_{n=1}^{\infty}$,' is omitted. This type of ellipsis is very common and makes for easier reading; the reader though should be on the lookout.

Next we do a similar reformulation for closed sets.

Proposition 2.13 *Let $W \subset X$. The following are equivalent:*

1) *W is closed.*
2) *If the sequence $(a_n)_{n=1}^{\infty}$ in W is convergent then its limit is in W.*

---

[2] We occasionally indulge in the entertaining sport of challenging the reader to find tacit applications of the axiom of choice.

Proof $\neg(1) \Rightarrow \neg(2)$. Suppose that W is not closed. Then $W^c := X \setminus W$ is not open. Hence there exists a convergent sequence $(b_n)_{n=1}^{\infty}$ whose limit y is in W^c but the sequence is not eventually in W^c (by proposition 2.12). But then the sequence $(b_n)_{n=1}^{\infty}$ has a subsequence that is wholly outside W^c, that is, the subsequence lies in W, but has the same limit y, which is outside W.

$(1) \Rightarrow (2)$. Let W be closed, let $(a_n)_{n=1}^{\infty}$ be a convergent sequence in W and let $b = \lim_{n \to \infty} a_n$. If $b \notin W$ then $b \in W^c$, which is open, and so $(a_n)_{n=1}^{\infty}$ is eventually in W^c. From this contradiction we conclude that $b \in W$.   □

Of the four clauses of the two preceding propositions, by far and away the most important is the rule that the limit of a convergent sequence in a closed set W lies in W. It is used over and over again, and possibly explains the predicate 'closed'. You cannot escape from a closed set by passing to the limit of a sequence.

2.2.6   Interior, Closure and Boundary

Let (X, d) be a metric space. Let $W \subset X$. We define three important operations that can be performed on W. They involve passing to new sets:

a) The *interior* of W, denoted by W°, is the union of all open sets U, such that $U \subset W$.
b) The *closure* of W, denoted by $\overline{W}$, is the intersection of all closed sets V, such that $W \subset V$.
c) The *boundary* of W, denoted by ∂W, is the set difference $\overline{W} \setminus W^{\circ}$.

It is clear by propositions 2.7 and 2.10 that W° is open and $\overline{W}$ is closed. Moreover W° is the maximum element in the set of all open sets that are subsets of W, ordered by inclusion. Similarly $\overline{W}$ is the minimum element in the set of all closed sets that include W, ordered by inclusion. It also follows that W is open if and only if $W = W^{\circ}$, and W is closed if and only if $W = \overline{W}$.

There is a simple criterion for a point to be in the interior and the proof is left to the reader:

Proposition 2.14 *A point x is in the interior of W if and only if W is a neighbourhood of x.*

To pass to the closure $\overline{W}$ of a given set W is one of the most important operations of metric space theory. In the next paragraphs we shall consider what points have to be joined to W to form its closure.

2.2.7 Limit Points of Sets

Let (X, d) be a metric space. If $W \subset X$ is not closed and $(a_n)_{n=1}^{\infty}$ is a convergent sequence in W, then the limit of the sequence is in the closure $\overline{W}$. This suggests that we might be able to form $\overline{W}$ by joining to W the limits of all convergent sequences in W. We shall explore this idea in the following pages, but first of all introduce the important notion of limit point of a set.

Definition Let $W \subset X$. A point $x \in X$ is called a *limit point* of W if every neighbourhood of x contains a point of W distinct from x.

An equivalent formulation of the condition in the definition is this: for every neighbourhood U of x we have

$$W \cap (U \setminus \{x\}) \neq \emptyset.$$

Proposition 2.15 *A point $x \in X$ is a limit point of W if and only if there exists a convergent sequence $(a_n)_{n=1}^{\infty}$ in $W \setminus \{x\}$ whose limit is x.*

Proof Suppose firstly that we have a sequence $(a_n)_{n=1}^{\infty}$, with the properties that $\lim_{n \to \infty} a_n = x$ and $a_n \in W \setminus \{x\}$ for all n. Let U be a neighbourhood of x. Then a_n is eventually in U but $a_n \neq x$. At the same time $a_n \in W$. We conclude that x is a limit point of W.

Suppose secondly that x is a limit point of W. For each $n \in \mathbb{N}_+$ there exists a point in $B_{1/n}(x) \cap W$ distinct from x. Choose one such and call it a_n. Then $a_n \in W \setminus \{x\}$ and $\lim_{n \to \infty} a_n = x$.

Note that the creation of the sequence $(a_n)_{n=1}^{\infty}$ involves an application of the axiom of choice to the family $(B_{1/n}(x) \cap W)_{n=1}^{\infty}$. $\qquad \square$

2.2.8 Characterisation of Closure by Limit Points

As always we work with a fixed metric space (X, d).

Proposition 2.16 *Let $W \subset X$. The closure $\overline{W}$ is the union of W and the set of all limit points of W.*

Proof Let $W^{\sim}$ denote the set of all limit points of W. We want to show that $\overline{W} = W \cup W^{\sim}$. There are two steps.

Step 1. Proof that $W \cup W^{\sim}$ is closed.

Suppose that $x \notin W \cup W^{\sim}$. We shall show that x has a neighbourhood that does not meet $W \cup W^{\sim}$.

Since x is neither in W nor is it a limit point of W, there exists $r > 0$, such that $B_r(x)$ contains no point of W. But then $B_r(x)$ cannot contain any point of $W^{\sim}$ either. For suppose, to the contrary, that $y \in B_r(x) \cap W^{\sim}$. Then $B_r(x)$ is a neighbourhood of the limit point y and therefore contains a point of W. This contradicts the choice of $B_r(x)$.

We see that $B_r(x)$ does not meet the set $W \cup W^\sim$. We conclude that $W \cup W^\sim$ is closed.

Step 2. Proof that $W \cup W^\sim$ is the minimum closed set that includes W.

Suppose that V is closed and that $W \subset V$. We shall show that $W^\sim \subset V$, by showing that

$$x \notin V \Rightarrow x \notin W^\sim.$$

Suppose that $x \notin V$. There exists $r > 0$, such that $B_r(x) \cap V = \emptyset$ (because V is closed). But then $B_r(x) \cap W = \emptyset$, so x not a limit point of W; that is, $x \notin W^\sim$. It follows that $W^\sim \subset V$ and so finally $W \cup W^\sim \subset V$.   □

The set of all limit points of a set W is often called the derived set of W. In the literature it is frequently denoted by W', whereas we used the more striking notation $W^\sim$. Our use of it will be very limited.

The next three propositions are trivial restatements of proposition 2.16, but provide different but useful pictures of the closure. They illustrate a feature of metric space theory, that the more important a notion, the more there are differing ways to talk about it.

Proposition 2.17 *A point x is in $\overline{W}$ if and only if every neighbourhood of x meets W.*

Proof This is an obvious restatement of the condition $x \in W \cup W^\sim$.   □

Proposition 2.18 *A set W is closed if and only if it contains all its limit points.*

Proof Since $\overline{W} = W \cup W^\sim$ we see that $W^\sim \subset W$ is equivalent to $\overline{W} = W$.   □

Proposition 2.19 *A point x is in $\overline{W}$ if and only if there exists a convergent sequence $(a_n)_{n=1}^\infty$ in W whose limit is x.*

Proof If $x \in \overline{W}$ then either x is in W or it is in $W^\sim$. In the first case the constant sequence $a_n = x$ in W converges to x. In the second case there exists a sequence $a_n \in W \setminus \{x\}$ whose limit is x.

For the converse, if $a_n \in W$ and $\lim_{n\to\infty} a_n = x$ then $x \in \overline{W}$.   □

As for the boundary, we have an appealing description, to stand beside the description of the closure in proposition 2.17 and the description of the interior in proposition 2.14:

Proposition 2.20 *A point x is in ∂W if and only if every neighbourhood of x contains a point of W and a point of W^c (the complement of W in X).*

Proof Recall that $\partial W = \overline{W} \setminus W^\circ$. We have that $x \in \overline{W}$ if and only if every neighbourhood of x contains a point in W; and $x \notin W^\circ$ if and only if every neighbourhood of x contains a point not in W.   □

2.2.9   Subspaces

We observed in section 1.2 that given a set that bears a metric, any non-empty subset of it can be made into a metric space in its own right and in a very natural way. In the context of multivariate analysis this means that any subset of $\mathbb{R}^n$ can be viewed as a metric space once we have agreed on a metric for $\mathbb{R}^n$. The most obvious choice is the Euclidean metric, but for applications other metrics may be simpler to work with. This is a vast array of metric spaces.

Let (X, d) be a metric space and let $M \subset X$. We form the metric space (M, d) by restricting d to M. We call (M, d) a *subspace* of (X, d).

Let $a \in M$. When we talk about an open ball with centre a we could mean an open ball in X, that is

$$\{x \in X : d(a, x) < r\},$$

or an open ball in M, that is

$$\{x \in M : d(a, x) < r\}.$$

If the intended meaning needs clarification we can use the notation $B_r^X(a)$ for the former and the notation $B_r^M(a)$ for the latter. For the closed balls we could use (though we rarely need to) $(B_r^M)^-(a)$ and $(B_r^X)^-(a)$. Then we have, given $a \in M \subset X$:

$$B_r^M(a) = B_r^X(a) \cap M, \quad (B_r^M)^-(a) = (B_r^X)^-(a) \cap M.$$

It is usual to speak about $B_r^M(a)$ as "the open ball with centre a and radius r relative to the set M" instead of the more formal "open ball with centre a and radius r in the metric space (M, d)".

In the same way we may say that a set $A \subset M$ that is open (respectively closed) in the metric space (M, d) is *open (respectively, closed) relative to the set M*. However, it is often more convenient to say, more simply, that *A is open (respectively closed) in M*.

Example Let $X = \mathbb{R}$ and $M = [0, 1]$. The open ball with centre 1 and radius $\frac{1}{2}$ relative to M is the set $]\frac{1}{2}, 1]$. It is open relative to M but not open in X.

2.2.10   Open and Closed Sets in a Subspace

Given a metric space (X, d) and a subset M of X we can describe the sets open relative to M, and the sets closed relative to M, in terms of the sets open, or closed in X, and without mentioning the metric.

Proposition 2.21 *Let (X, d) be a metric space, let $M \subset X$ and $A \subset M$. Then:*

1) *The set A is open relative to M if and only if there exists a set $U \subset X$, such that U is open in X and $A = M \cap U$.*
2) *The set A is closed relative to M if and only if there exists a set $V \subset X$, such that V is closed in X and $A = M \cap V$.*

Proof

1) Let A be open in (M, d). For each $x \in A$ there exists $r_x > 0$, such that

$$B_{r_x}^{M}(x) \subset A.$$

Set

$$U := \bigcup_{x \in A} B_{r_x}^{X}(x).$$

Then U is open in (X, d) and

$$M \cap U = \bigcup_{x \in A} M \cap B_{r_x}^{X}(x) = \bigcup_{x \in A} B_{r_x}^{M}(x) = A.$$

Conversely, suppose that $A = M \cap U$, where U is open in X. Let $x \in A$. Then $x \in M$ and also $x \in U$. Hence there exists $r > 0$, such that

$$B_{r}^{X}(x) \subset U.$$

But then

$$B_{r}^{M}(x) = B_{r}^{X}(x) \cap M \subset U \cap M = A,$$

that is, $B_{r}^{M}(x) \subset A$ and we conclude that A is open relative to M.

2) Suppose that $A \subset M$ and is closed relative to M. Then $M \setminus A$ is open relative to M. Hence, by item 1, there exists U, open in X, such that

$$M \setminus A = M \cap U.$$

We perform the Boolean calculation (with complementation in X):

$$A = M \setminus (M \setminus A) = M \setminus (M \cap U) = M \setminus U = M \cap U^{c}$$

and observe that U^{c} is closed in X.

Suppose, conversely, that $A \subset M$ and there exists V, closed in X, such that $A = M \cap V$. Then

$$M \setminus A = M \setminus (M \cap V) = M \cap V^c.$$

Now V^c is open in X and by item 1 we find that $M \setminus A$ is open relative to M; so that A is closed relative to M.   □

2.2.11 Exercises

1. Show that a one-point set $\{a\}$ in a metric space (X, d) is always closed.
2. A point a in a metric space (X, d) is said to be isolated if the set $\{a\}$ is open. Show that given a set X it is possible to define a metric on X so that all points are isolated in (X, d).
3. Show that open intervals $]a, b[$ are open sets and closed intervals $[a, b]$ are closed sets in the metric space $\mathbb{R}$ with its usual metric $|s - t|$.
4. Show that Cantor's middle thirds set B (section 1.3) is closed in $\mathbb{R}$.
5. Show that every point of Cantor's middle thirds set B is a limit point of B.
6. Find an example of a metric space (X, d) and a countable infinite family of closed sets $(A_n)_{n=1}^{\infty}$ in X, such that their union $\bigcup_{n=1}^{\infty} A_n$ is not closed.
7. Let A be a non-empty set in a metric space X and let $v \in X$. Show that $d(v, A) = d(v, \overline{A})$.
8. Let (X, d) be a metric space, $a \in X$ and $r > 0$. We know that $B_r(a)$ (open ball) is an open set and $B_r^-(a)$ (closed ball) is a closed set.

   a) Show that

   $$B_r(a) \subset \overline{B_r(a)} \subset B_r^-(a).$$

   b) Show that the set $B_r^-(a) \setminus B_r(a)$, the "sphere" of radius r, is closed.

9. It is possible to have $\overline{B_r(a)} \neq B_r^-(a)$. In such a case the boundary of $B_r(a)$ is not the same as the set difference $B_r^-(a) \setminus B_r(a)$, the "sphere" of radius r. Examples are not necessarily exotic. Here are two:

   a) Let $X = \mathbb{N}$, viewed as a subspace of $\mathbb{R}$ with the usual metric. Show that in X we have

   $$B_1(0) = \overline{B_1(0)} \neq B_1^-(0).$$

   b) Let $M \subset \mathbb{R}^2$ be the set

   $$\{(x, 0) : 0 \leq x \leq 1\} \cup \{(1, y) : 0 \leq y \leq 1\}$$

and view it as a subspace of $\mathbb{R}^2$ with the metric induced by the norm $\| \cdot \|_\infty$. Compute the balls $B_1^M\big((0,0)\big)$ and $(B_1^M)^-\big((0,0)\big)$ and show that the closure in M of the former is not the same as the latter.

c) What is the boundary of the set $B_1^M\big((0,0)\big)$ in the metric space M defined in item (b)?

10. Prove the following often used rules about closure and interior:

   a) $\overline{\overline{A}} = \overline{A}$.
   b) $A^{\circ\circ} = A^\circ$.
   c) If $A \subset B$ then $\overline{A} \subset \overline{B}$.
   d) If $A \subset B$ then $A^\circ \subset B^\circ$.
   e) $\overline{A \cup B} = \overline{A} \cup \overline{B}$.
   f) $\overline{A \cap B} \subset \overline{A} \cap \overline{B}$. Show by an example that the inclusion can be proper.

11. Find the boundary of A for the following subsets of $\mathbb{R}$:

   a) $A = \mathbb{R}$.
   b) $A = \mathbb{Q}$.
   c) $A = \mathbb{R} \setminus \mathbb{Q}$.
   d) A is Cantor's middle thirds set.

12. Let (X, d) be a metric space and let $A \subset X$. Prove the following claims about the boundary of ∂A of A:

   a) ∂A is closed.
   b) $\partial A = \overline{A} \cap \overline{A^c}$.
   c) $\partial(A^c) = \partial A$.
   d) The set X is partitioned into three pairwise disjoint sets by

$$X = A^\circ \cup (A^c)^\circ \cup \partial A.$$

13. Some more rules you can prove:

   a) $\overline{A} = \big((A^c)^\circ\big)^c$.
   b) $A^\circ = (\overline{A^c})^c$.

14. Let A be a set in a metric space X.

   a) Show that if A is closed then $(\partial A)^\circ = \emptyset$.
   b) Show that $\partial\partial\partial A = \partial\partial A$.

15. Let A and B be non-empty sets in a metric space X. Show that $d_H(A, B) = 0$ (d_H being the Hausdorff metric, see section 1.2) if and only if $\overline{A} = \overline{B}$. So the Hausdorff metric, which in general is only a pseudometric, is a genuine metric (though one that can take the value ∞) when restricted to the non-empty closed sets.

16. Show that if X is a normed vector space then $\partial B_r(a) = B_r^-(a) \setminus B_r(a)$.

17. Let (X, d) be an ultrametric space (1.2 exercises 1.2.6 and 1.2.6, and 2.1 exercise 9). Prove the following counterintuitive properties:

   a) The open ball $B_r(a)$ is a closed set (as well as being open, as in all metric spaces).
   b) The "sphere" of radius r, the set

$$S_r(a) := \{x \in X : d(x, a) = r\} = B_r^-(a) \setminus B_r(a),$$

   is open (as well as being closed, as in all metric spaces).
   c) The closed ball $B_r^-(a)$ is an open set (as well as being closed, as in all metric spaces).

18. For this exercise you will need to know about equivalence relations and partitions. Let (X, d) be an ultrametric space.

   a) Show that for each fixed $r > 0$ the relations $d(x, y) < r$ and $d(x, y) \leq r$ are equivalence relations.
   b) Let $r > 0$. Show that the space X is partitioned by the set of all open balls of radius r, and also by the set of all closed balls of radius r.
   c) Let $0 < s < r$. Show that the open ball $B_r(a)$ is partitioned by the set of all open balls of radius s that meet it.
   d) Let $\mathbb{Z}$ be equipped with the dyadic metric. Describe the partition of $\mathbb{Z}$ by the set of all open balls of radius $\frac{1}{2}$, and also by the set of all closed balls of radius $\frac{1}{2}$.

19. Let $C_0(\mathbb{R})$ denote the set of all continuous functions $f : \mathbb{R} \to \mathbb{R}$ such that $\lim_{|x| \to \infty} f(x) = 0$. Functions in $C_0(\mathbb{R})$ are sometimes called continuous functions that vanish at infinity. Show that $C_0(\mathbb{R})$ is closed in $C_B(\mathbb{R})$, the latter being the space of bounded continuous functions with the supremum norm $\|f\|_\infty$ (section 1.4).

20. Let X be a vector space over $\mathbb{R}$. Recall the notion of convex set (1.2 exercise 1.2.6). A set $A \subset X$ is said to be convex, if, for all $x, y \in A$ and all $t \in [0, 1]$, the vector $(1 - t)x + ty$ is in A.

   a) Let $(A_i)_{i \in I}$ be a family of convex subsets of X. Show that $\bigcap_{i \in I} A_i$ is convex.
   b) Let $B \subset X$. Show that there exists a minimum (in the sense of set inclusion) convex set that includes B. It is called the *convex hull* of B.
   c) Show that the convex hull of B consists of all vectors expressible in the form $\sum_{k=1}^{n} t_k x_k$, for an integer n, vectors $x_k \in B$ and real numbers $t_k \in [0, 1]$ $(k = 1, \ldots, n)$, satisfying $\sum_{k=1}^{n} t_k = 1$.
   d) Suppose that X is a normed space. Let A be a convex subset of X. Show that $\overline{A}$ (the closure of A) is convex.
   e) Let X be a normed space. Let $B \subset X$. Show that there exists a minimum closed convex set including B. It is called the *closed convex hull* of B.
   f) Show that the closed convex hull of B is the closure of the convex hull of B.

g) Let $X = \ell^2$, let $e^{(n)} = (a_k)_{k=1}^{\infty}$ where $a_n = 1$ and $a_k = 0$ for $k \neq n$. Let

$$A = \{0\} \cup \{e^{(n)} : n \in \mathbb{N}_+\} \subset \ell^2.$$

Show that A is closed in X, identify the convex hull and the closed convex hull of A and show that they are distinct. Thus the convex hull of a closed set need not be closed.

h) Show that the convex hull of an open set is open.

i) You don't have to be in an infinite dimensional space to find an example of a closed set whose convex hull is not closed. Consider the curve $y = e^{-|x|}$ in the plane $\mathbb{R}^2$.

21. Let (X, d) be a metric space and let M be a subset of X. Show the following:

   a) If M is open, then for every subset $A \subset M$: A is open relative to M if and only if A is open in X.

   b) If M is closed, then for every subset $A \subset M$: A is closed relative to M if and only if A closed in X.

22. Let (X, d) be a metric space.

   a) Let $U \subset A \subset B \subset X$. Show that if U is open (respectively, closed) relative to B then it is open (respectively, closed) relative to A.

   b) Let $U \subset B \subset X$ and $A \subset B \subset X$. Show that if U is open (respectively, closed) relative to B then $U \cap A$ is open (respectively, closed) relative to A.

   Note. The simple rules in this and the preceding exercise are frequently used in dealing with subspaces.

23. Let (X, d) be a metric space. Let A and B be non-empty subsets of X. Suppose that $U \subset A \cap B$ and that U is open (respectively, closed) relative to A and open (respectively, closed) relative to B. Show that U is open (respectively, closed) relative to $A \cup B$.
   Hint. If A, B, M, N and U are subsets of X, such that

$$U = M \cap A = N \cap B.$$

   Then

$$U = (M \cap N) \cap (A \cup B).$$

24. Let X be a metric space and let A be a non-empty subset of X. For each $r > 0$ the set

$$N_r(A) := \{x \in X : d(x, A) < r\}$$

is called the open r-neighbourhood of A. We expressly require that A is non-empty as we cannot define $d(x, \emptyset)$. Show that:

a) $N_r(A) = \bigcup_{x \in A} B_r(x)$.

b) $N_r(A)$ is open.

c) $\overline{N_r(A)} \subset \{x \in X : d(x, A) \leq r\}$.

d) If $0 < r < s$ then $\overline{N_r(A)} \subset N_s(A)$.

25. Let M and N be closed sets in a metric space X. Show that there exist open sets U and V, such that $M \subset U$, $N \subset V$ and $U \cap V = \emptyset$.

   Hint. Use 2.1 exercise 8. Note that the infimum of $d(x, y)$ for $x \in M$ and $y \in N$ can be 0, so it may not work to take $U = N_r(M)$ and $V = N_r(N)$ for a suitably small r (in the notation of exercise 24).

26. In a metric space X, let M be a closed set and U an open set, such that $M \subset U$. Show that there exists an open set V, such that

$$M \subset V \subset \overline{V} \subset U.$$

27. In a metric space X let $(U_k)_{k=1}^n$ be a finite family of open sets, such that

$$X = \bigcup_{k=1}^n U_k.$$

Show that there exist open sets V_k, $(k = 1, \ldots, n)$, such that

$$V_k \subset \overline{V}_k \subset U_k, \quad (k = 1, \ldots, n)$$

and

$$X = \bigcup_{k=1}^n V_k.$$

Hint. The premises imply that

$$\left(\bigcup_{k=2}^n U_k \right)^c \subset U_1.$$

Apply the result of the previous exercise and continue by induction.

28. Prove the following much used structural property of open sets in $\mathbb{R}$:

   An open set $A \subset \mathbb{R}$ is a countable union of pairwise disjoint open intervals.

Hints for a solution:

a) Define a relation on A by stipulating: $x \sim y$ if and only if there exists an open interval I, such that $x \in I$, $y \in I$ and $I \subset A$. Show that $x \sim y$ is an equivalence relation on A.

b) Show that the equivalence classes are open intervals.

c) Show that the set of all equivalence classes is countable.

2.3 Continuous Mappings

The utility of metric spaces is due to much more than their ubiquity. It is due, first and foremost, to the fact that the continuity concept makes sense for mappings between metric spaces.

2.3.1 Defining Continuity

We consider two metric spaces (X, d_X) and (Y, d_Y).

Definition Let $f : X \to Y$ and let $a \in X$. We say that f is continuous at the point a if the following condition is satisfied:

For all $\varepsilon > 0$ there exists $\delta > 0$, such that $d_Y(f(x), f(a)) < \varepsilon$ for all $x \in X$ that satisfy $d_X(x, a) < \delta$.

In this definition we have followed closely the usual (ε, δ)-definition of continuity of fundamental analysis, simply replacing absolute values by the metrics. However, we can also express the condition for continuity at a with open balls, without really changing anything, and achieve some economy of thought without sacrificing clarity:

For all $\varepsilon > 0$ there exists $\delta > 0$, such that $f(B_\delta(a)) \subset B_\varepsilon(f(a))$.

We should understand by the context whether $B_r(\cdot)$ refers to X or Y. In future we will often omit the subscripts in 'd_X' and 'd_Y', and denote both metrics by d, if there seems to be no cause for misunderstanding.

Definition A mapping $f : X \to Y$ is said to be continuous if it is continuous at each $a \in X$.

In fundamental analysis it is necessary to treat functions whose domains are subsets of $\mathbb{R}$, but not necessarily all of $\mathbb{R}$. This would include nearly all the elementary functions of analysis. In the context of metric spaces one might like to consider a function $f : A \to Y$, where $A \subset X$, given the metric spaces (X, d_X) and

(Y, d_Y). When we say that such a function is continuous, we shall always mean that it is continuous when A is regarded as a subspace of X, that is, when A is viewed as metric space (A, d_X) in its own right. This is important to remember when we study the inverse of f, and its continuity. In the situation that f is injective it has an inverse whose domain is $f(X)$, often a proper subset of Y. When we ask whether the inverse is continuous we must view its domain as a subspace of Y.

2.3.2 New Views of Continuity

We continue to assume two metric spaces (X, d_X) and (Y, d_Y).

Proposition 2.22 *Let $f : X \to Y$. Then f is continuous at $a \in X$ if and only if, for every neighbourhood U of $f(a)$, the inverse image $f^{-1}(U)$ is a neighbourhood of a.*

Proof Suppose first that $f^{-1}(U)$ is a neighbourhood of a whenever U is a neighbourhood $f(a)$. Let $\varepsilon > 0$. We may let U be the neighbourhood $B_\varepsilon(f(a))$ of $f(a)$. Now $f^{-1}(B_\varepsilon(f(a)))$ is a neighbourhood of a and so there exists $\delta > 0$, such that

$$B_\delta(a) \subset f^{-1}(B_\varepsilon(f(a))).$$

But then

$$f(B_\delta(a)) \subset B_\varepsilon(f(a)).$$

We conclude that f is continuous at a.

Suppose next that f is continuous at a. We want to show that $f^{-1}(U)$ is a neighbourhood of a whenever U is a neighbourhood $f(a)$. So let U be a neighbourhood of $f(a)$. There exists $\varepsilon > 0$, such that

$$B_\varepsilon(f(a)) \subset U,$$

and then $\delta > 0$, such that

$$f(B_\delta(a)) \subset B_\varepsilon(f(a)).$$

But then

$$B_\delta(a) \subset f^{-1}\big(B_\varepsilon(f(a))\big) \subset f^{-1}(U).$$

We conclude that $f^{-1}(U)$ is a neighbourhood of a. □

Proposition 2.23 *Let* $f : X \to Y$. *The following conditions are equivalent:*

1) f *is continuous.*
2) $f^{-1}(U)$ *is open in X whenever U is open in Y.*
3) $f^{-1}(U)$ *is closed in X whenever U is closed Y.*

Proof (2) $\Leftrightarrow$ (3). An obvious consequence of the rule

$$f^{-1}(U^c) = (f^{-1}(U))^c$$

and the definition of closed set.

(1) $\Rightarrow$ (2). Let f be continuous and U open in Y. Let $a \in f^{-1}(U)$. Then $f(a) \in U$ and U is a neighbourhood of $f(a)$. Then $f^{-1}(U)$ is a neighbourhood of a (by proposition 2.22). We conclude that $f^{-1}(U)$ is open.

(2) $\Rightarrow$ (1). Given condition 2 we let $a \in X$. Let V be a neighbourhood of $f(a)$. We shall show that $f^{-1}(V)$ is a neighbourhood of a. There exists an open neighbourhood U of $f(a)$ (for example an open ball), such that $U \subset V$. Then $f^{-1}(U)$ is open, contains a and is therefore a neighbourhood of a. But $f^{-1}(U) \subset f^{-1}(V)$ so that the latter is also a neighbourhood of a.   □

The two preceding propositions exhibit a striking progression in which quantifiers are repeatedly reduced in number. The (ε, δ)-definition of continuity at a point a contains three quantifiers, having the form

$$(\forall \varepsilon)(\exists \delta)(\forall x)(\ldots).$$

The rephrased definition using balls has two, being of the form

$$(\forall \varepsilon)(\exists \delta)(\ldots).$$

The third quantifier has been absorbed into the subset relation. Proposition 2.22 presents a condition for continuity at a that uses only one quantifier,

$$(\forall U)(\ldots);$$

the notion of neighbourhood has absorbed another one.

When we examine the definition of continuity of a mapping $f : X \to Y$ in its (ε, δ)-form, as in fundamental analysis, we see a row of four quantifiers, something like

$$(\forall a)(\forall \varepsilon)(\exists \delta)(\forall x)(\ldots).$$

The same thing is expressed by proposition 2.23 with only one implied quantifier,

$$(\forall U)(\ldots).$$

Comprehending a row of nested quantifiers can be a challenge. The clarity and simplicity of proposition 2.23, even if achieved by defining predicates and relations ('open', 'subset' *etc.*) that so-to-speak sweep the quantifiers under the carpet, is truly remarkable.

2.3.3   Limits of Functions

In treatments of fundamental analysis the notion of the limit of a function, $\lim_{x \to a} f(x)$, plays a basic role. Often it is introduced before continuity, the latter being defined by the requirement that $f(a) = \lim_{x \to a} f(x)$. In metric space theory its role is less prominent than in analysis.

Let (X, d_X) and (Y, d_Y) be metric spaces, let $M \subset X$ and let $f : M \to Y$. Let a be a limit point of M in X. We do not require that a is in M. We can define the notion $\lim_{x \to a} f(x)$ as follows. Let $b \in Y$. We say that f has the limit b at the point a, and write

$$\lim_{x \to a} f(x) = b$$

if the following condition is satisfied:

For all $\varepsilon > 0$ there exists $\delta > 0$, such that $d_Y(f(x), b) < \varepsilon$ for all $x \in M$ that satisfy $0 < d_X(x, a) < \delta$.

We do not define the limit at a if a is not a limit point of M. An alternative form of the condition that does not mention the metric is:

For each neighbourhood V of b there exists a neighbourhood U of a, such that $f(U \cap M \setminus \{a\}) \subset V$.

The following proposition relates limits and continuity in the manner familiar from fundamental analysis. The proof is left to the reader.

Proposition 2.24 *Let $f : X \to Y$ and let a be a limit point of X. Then f is continuous at a if and only if $\lim_{x \to a} f(x) = f(a)$.*

We can also frame a version of Cauchy's principle of convergence for limits of functions, that is often useful in analysis.

Proposition 2.25 *Let X and Y be metric spaces and assume that Y is complete. Let $M \subset X$, let $f : M \to Y$ and let a be a limit point of M. Then the limit $\lim_{x \to a} f(x)$ exists if and only if the following condition is satisfied: for all $\varepsilon > 0$ there exists a neighbourhood U of a, such that, for all x and y in $M \cap U \setminus \{a\}$ we have $d(f(x), f(y)) < \varepsilon$.*

Proof Suppose that the condition is satisfied. Let $(c_n)_{n=1}^{\infty}$ be a sequence in $M \setminus \{a\}$ that converges to a (see proposition 2.15). It is easy to see that the sequence

$(f(c_n))_{n=1}^\infty$ is a Cauchy sequence in Y, hence is convergent. Let its limit be b. We proceed to show that $\lim_{x \to a} f(x) = b$.

Let $\varepsilon > 0$. There exists a neighbourhood U of a, such that for all x and y in $M \cap U \setminus \{a\}$ we have $d(f(x), f(y)) < \frac{1}{2}\varepsilon$. We can find m, such that both $c_m \in U$ and also $d(f(c_m), b) < \frac{1}{2}\varepsilon$. Now if $x \in M \cap U \setminus \{a\}$ we have $d(f(x), f(c_m)) < \frac{1}{2}\varepsilon$, and hence $d(f(x), b) < \varepsilon$ by the triangle inequality. We conclude that $\lim_{x \to a} f(x) = b$.

Conversely, suppose that the limit $\lim_{x \to a} f(x)$ exists and equals b, say. Let $\varepsilon > 0$. There exists a neighbourhood U of a, such that for all $x \in M \cap U \setminus \{a\}$ we have $d(f(x), b) < \frac{1}{2}\varepsilon$. Now if x and y are in $M \cap U \setminus \{a\}$ we have $d(f(x), f(y)) < \varepsilon$, by the triangle inequality. $\square$

The form of Cauchy's principle just enunciated can often be exploited in combination with the concept of *oscillation of a function*.

Definition Let X and Y be metric spaces, let $f : X \to Y$ and let A be a non-empty subset of X. The oscillation of f in A is defined as

$$\Omega_f(A) = \sup_{x, y \in A} d\big(f(x), f(y)\big).$$

We allow the value ∞ if the set of real numbers for which the supremum is required is unbounded.

Let $a \in X$. The *point oscillation* of f at the point a is defined as

$$\omega_f(a) = \inf\{\Omega_f(U) : U \text{ a neighbourhood of } a\}.$$

The function ω_f maps X to the set $\mathbb{R} \cup \{\infty\}$. We call $\omega_f(a)$ the point oscillation to distinguish it from the oscillation in the set $\{a\}$, which is of no interest, being always 0.

The proofs of the remaining propositions of this section are left to the reader as exercises. Throughout them we assume that X and Y are metric spaces and $f : X \to Y$.

Proposition 2.26 *Let $a \in X$. Then f is continuous at a if and only if $\omega_f(a) = 0$.*

Proposition 2.27 *Let a be limit point of X and suppose that the limit $\lim_{x \to a} f(x)$ exists. Then*

$$\lim_{x \to a} \omega_f(x) = 0.$$

Proposition 2.28 *Let Y be complete and let a be a limit point of X. Then the limit $\lim_{x \to a} f(x)$ exists if and only if*

$$\lim_{r \to 0+} \Omega_f(B_r(a) \setminus \{a\}) = 0.$$

Exercise Prove these propositions. Show also that the converse of proposition 2.27 is false, even if Y is complete.

2.3.4 Characterising Continuity by Sequences

The main importance of the following proposition lies in its permitting the interchange of applying a function and taking a limit. This makes it a useful tool in approximation theory, in particular when one computes a quantity by iterations. It is sometimes useful for verifying that a function is continuous.

We assume that X and Y are metric spaces. Each is therefore equipped with a metric, to which we need not refer.

Proposition 2.29 *Let $f : X \to Y$. The following conditions are equivalent:*

1) f is continuous.
2) If $x \in X$, and if $(a_n)_{n=1}^{\infty}$ is a sequence in X such that $\lim_{n\to\infty} a_n = x$, then $\lim_{n\to\infty} f(a_n) = f(x)$.

Proof (1) $\Rightarrow$ (2). Assume that f is continuous, that $x \in X$ and that $(a_n)_{n=1}^{\infty}$ is a sequence whose limit is x. Let V be a neighbourhood of $f(x)$. Then $f^{-1}(V)$ is a neighbourhood of x and therefore, by proposition 2.3, the sequence $(a_n)_{n=1}^{\infty}$ is eventually in $f^{-1}(V)$. But this says that $(f(a_n))_{n=1}^{\infty}$ is eventually in V; in other words we have $\lim_{n\to\infty} f(a_n) = f(x)$.

$\neg$(1) $\Rightarrow$ $\neg$(2). Assume that f is not continuous. Then, by proposition 2.23, there exists a set V, open in Y, such that $f^{-1}(V)$ is not open in X. It follows, by proposition 2.12, that there exists $x \in f^{-1}(V)$ and a sequence $(a_n)_{n=1}^{\infty}$ in $X \setminus f^{-1}(V)$ whose limit is x. But then $f(a_n) \notin V$, and since V is open and $f(x) \in V$ it is impossible that $f(a_n) \to f(x)$. $\square$

2.3.5 Lipschitz Mappings

Let (X, d_X) and (Y, d_Y) be metric spaces.

Proposition 2.30 *Let $f : X \to Y$ and assume that there exists K, such that for all x and x' in X we have*

$$d_Y\big(f(x), f(x')\big) \leq K d_X(x, x').$$

Then f is continuous.

Proof The condition may be rephrased as follows: for all $a \in X$ we have $f(B_{\varepsilon/K}(a)) \subset B_{\varepsilon}(f(a))$. Therefore f is continuous. $\square$

The condition here imposed on f is called a Lipschitz condition and mappings that satisfy a Lipschitz condition are aptly called Lipschitz mappings. Such a condition is by no means necessary for continuity. In the case when $K < 1$ we call f a strict contraction mapping. They have important applications to proving the existence of solutions to equations (chapter 6).

2.3.6   Examples of Continuous Functions

All the functions proved to be continuous in fundamental analysis (essentially, this is the scope of single variable calculus) are continuous in the sense of this chapter. We assume of course that $\mathbb{R}$ has its usual metric $|s - t|$ and there is no need to repeat the usual proofs. But what about metric spaces in general? Can we always find some continuous functions?

We shall show that every metric space that has a proper, closed, and non-empty subset is the domain of a non-constant continuous function. Let (X, d) be a metric space. Recall the definition of the distance from a point to a set. Let $A \subset X$, and assume that $A \neq \emptyset$. For each $x \in X$ we set

$$d(x, A) = \inf\{d(x, y) : y \in A\}.$$

The function $d(x, A)$ gives us yet another way to view the situation where x is in the closure of A, to stand beside propositions 2.16, 2.17, 2.18 and 2.31.

Proposition 2.31 *Let A be a non-empty subset of X. Then $x \in \overline{A}$ if and only if $d(x, A) = 0$.*

Proof By proposition 2.17, we know that $x \in \overline{A}$ if and only if every neighbourhood of x contains a point of A. By the definition of neighbourhood this is equivalent to saying that for all $\varepsilon > 0$ the ball $B_\varepsilon(x)$ contains a point of A; which is to say that for all $\varepsilon > 0$ there exists $y \in A$ such that $d(x, y) < \varepsilon$; or equivalently, $\inf_{y \in A} d(x, y) = 0$, as we wished to prove.   □

Exercise The infimum of a subset of $[0, \infty[$ is 0 if and only if it includes a sequence with limit 0. Use this to give a another proof of proposition 2.31 using the characterisation of closure by convergent sequences.

We recall the formula

$$|d(x, A) - d(y, A)| \leq d(x, y).$$

Thus, taking $\mathbb{R}$ to have the usual metric (an assumption which we won't bother to state in future), the function $f(x) = d(x, A)$ is continuous; in fact it has Lipschitz constant 1. Its zero-set is the closure of A. So provided A is not empty and $\overline{A}$ is not all of X, we obtain a non-constant continuous function on X. Every closed set is the zero-set of a continuous real-valued function.

Building on the function $d(x, A)$ we can produce further examples of continuous functions with domain X. The possibilities are limitless. For example the function $f(d(x, A))$, where f is a continuous function of a real variable (an example of composition of continuous functions—see exercise 1); or involving several sets, as in $d(x, A) - d(x, B)$.

The function $d(x, y)$ is a function of two variables. Its domain is $X \times X$. To make sense of the question "Is d continuous?" we have to assign a metric to $X \times X$. We treat the case of two spaces here in order to give a second example of a continuous mapping defined on an abstract metric space. The study of products will be taken up in greater length later.

Definition Let (X, d_X) and (Y, d_Y) be metric spaces. Define the metric D on $X \times Y$ (the product metric) by

$$D\big((x, y), (x', y')\big) = \max\big(d_X(x, x'), d_Y(y, y')\big)$$

Other metrics on $X \times Y$ can be obtained by fixing a number $p \geq 1$ and setting

$$D_p\big((x, y), (x', y')\big) = \big(d_X(x, x')^p + d_Y(y, y')^p\big)^{1/p}.$$

Generally we shall use D, defined first, as the default metric on $X \times Y$. For most purposes it does not matter which of these metrics is used. It is left to the reader to show that these are indeed metrics.

Now we can show that the function $d : X \times X \to \mathbb{R}$ is continuous. We have

$$|d(x, y) - d(x', y')| \leq |d(x, y) - d(x, y')| + |d(x, y') - d(x', y')|$$

$$\leq d(y, y') + d(x, x') \leq 2D\big((x, y), (x', y')\big).$$

So d satisfies a Lipschitz condition with constant 2.

2.3.7 Exercises

1. Let X, Y, Z be metric spaces (each assigned a metric of course, but you may not need to mention them). Let $f : X \to Y$ and $g : Y \to Z$ be continuous mappings. Show that $g \circ f$ is a continuous mapping from X to Z.
2. Suppose one tried to define the limit $\lim_{x \to a} f(x)$, where $f : M \to Y, M \subset X$, and a is *not a limit point* of M, using the same definition as was given in the text. What would the limit be?
3. Let X, Y, Z be metric spaces, with metrics d_X, d_Y and d_Z respectively. Define the projections (also called coordinate functions)

$$\pi_1 : X \times Y \to X, \qquad \pi_2 : X \times Y \to Y$$

by

$$\pi_1(x, y) = x, \qquad \pi_2(x, y) = y.$$

a) Show that π_1 and π_2 are continuous mappings.
b) Let $f : Z \to X \times Y$ where $X \times Y$ has the default product metric. Show that f is continuous if and only if the mappings $\pi_1 \circ f$ and $\pi_2 \circ f$ (the *coordinates* of f) are both continuous.

4. Let X, Y, Z be metric spaces. Let $f : X \times Y \to Z$. Given a point $(a, b) \in X \times Y$ we define the mappings (the *partial mappings* of f)

$$f_a : Y \to Z, \quad f_a(y) = f(a, y), \quad (y \in Y),$$

and

$$f^b : X \to Z, \quad f^b(x) = f(x, b), \quad (x \in X).$$

a) Show that if f is continuous then f_a and f^b are continuous for all a and b.
b) Find an example of a function $f : \mathbb{R} \times \mathbb{R} \to \mathbb{R}$, such that f_a and f^b are continuous for all $a \in \mathbb{R}$ and $b \in \mathbb{R}$, but f is not continuous.

5. Let X and Y be metric spaces and let $f : X \to Y$. Show that continuity of f is equivalent to each of the following conditions:

a) For all $A \subset X$: $f(\overline{A}) \subset \overline{f(A)}$.
b) For all $B \subset Y$: $f^{-1}(B^\circ) \subset f^{-1}(B)^\circ$.
c) For all $B \subset Y$: $f^{-1}(B) \subset f^{-1}(\overline{B})$.

In each case find an example of spaces, a continuous mapping and sets such that the inclusion is proper.

6. Show that the set $A \subset \mathbb{R}^3$ defined by

$$A = \left\{ (x, y, z) \in \mathbb{R}^3 : \forall (p \geq 1) \left(\left(|x|^p + |y|^p + |z|^p \right)^{1/p} \leq 1 + 1/p \right) \right\}$$

is closed. (The solution is short. Use your knowledge of analysis.)

7. Prove the "rising sun" lemma. Let $f : \mathbb{R} \to \mathbb{R}$ be continuous. Call a point $x \in \mathbb{R}$ a *shadow point* of f if there exists $y > x$ such that $f(y) > f(x)$.

a) Show that the set of all shadow points is open.
   Hint. A smart way is to observe that the set of shadow points is the set

$$\bigcup_{y \in \mathbb{R}} \left( ]-\infty, y[\cap f^{-1}(]-\infty, f(y)[) \right).$$

b) By 2.2 exercise 28 the set of all shadow points is the union of a countable, pairwise disjoint family of open intervals. Suppose that $]a, b[$ is a bounded member of this family (assuming there is one). Show that $f(a) = f(b)$.

   Hint. Let $a < x < b$. Bear in mind that b is not a shadow point, whereas every point of the interval $[x, b[$ is. Hence show that the maximum of f with respect to the interval $[x, b]$ is attained solely at b.

8. Sketch the graph of the function $d(x, A)$, where $A \subset \mathbb{R}$, in the following cases:

   a) $A = \{0\}$.
   b) $A = \{0, 1, 2\}$.
   c) $A =]-\infty, -1] \cup [1, \infty[$.

9. Let X and Y be metric spaces and let $f : X \to \mathbb{R}$ and $g : Y \to \mathbb{R}$ be continuous functions. Prove the following?

   a) The set $\{x \in X : f(x) < g(x)\}$ is open.
   b) The set $\{x \in X : f(x) \le g(x)\}$ is closed.
   c) The set $\{x \in X : f(x) = g(x)\}$ is closed.

10. Give a quick solution to 2.2 exercise 25 by exploiting the continuity of the function $d(x, M) - d(x, N)$ and proposition 2.31.

11. Let (X, d) be a metric space and let $A \subset X$. Recall that the characteristic function (also called indicator function) of A is the function $\chi_A : X \to \mathbb{R}$, such that $\chi_A(x) = 1$ if $x \in A$ and $\chi_A(x) = 0$ if $x \notin A$.

   a) Show that for each $a \in X$ the function χ_A is continuous at a if and only if $a \in A° \cup (A^c)°$.
   b) Show that the function χ_A is discontinuous at a if and only if a is in the boundary ∂A of A.

12. It is a familiar fact that an open interval of real numbers can be expressed as a countable union of an increasing sequence of bounded closed intervals. In this exercise we explore whether an open set in a metric space can be expressed as the union of an increasing sequence of bounded closed sets.

   Let X be a metric space and let $A \subset X$ be open. Fix some point $a \in X$. For each positive integer n we let

   $$E_n = B_n(a) \cap \{x \in X : d(x, A^c) > \tfrac{1}{n}\},$$
   $$F_n = B_n^-(a) \cap \{x \in X : d(x, A^c) \ge \tfrac{1}{n}\}.$$

   For this to make sense the set A^c must be non-empty. In the case that $A = X$ we set $E_n = B_n(a)$ and $F_n = B_n^-(a)$.

   Show that E_n is open and bounded, F_n is closed and bounded, that

   $$E_n \subset F_n \subset E_{n+1}$$

for each n, and that

$$A = \bigcup_{n=1}^{\infty} E_n = \bigcup_{n=1}^{\infty} F_n.$$

13. Let X be a metric space and let $(A_i)_{i \in I}$ be a family of subsets of X such that $X = \bigcup_{i \in I} A_i$. Let Y be another metric space and suppose that for each $i \in I$ we have a continuous mapping $f_i : A_i \to Y$. Suppose further that the family of mappings $(f_i)_{i \in I}$ satisfies

$$f_i|_{A_i \cap A_j} = f_j|_{A_i \cap A_j}$$

whenever $A_i \cap A_j \neq \emptyset$. Then, by set theory, there exists a unique mapping $g : X \to Y$, such that

$$g|_{A_i} = f_i$$

for each $i \in I$.

   Prove that the mapping g is continuous in each of the following cases:

   a) Each set A_i is open.
   b) Each set A_i is closed and the index set I is finite.

   Give an example to show that continuity of g may fail if each set A_i is closed and I is infinite.

14. Let X be a metric space and let A be closed in X. Suppose that the function $f : A \to \mathbb{R}$ is continuous and that $f(x) = 0$ for all $x \in \partial A$. Show that f can be extended to a continuous function $g : X \to \mathbb{R}$, by setting $g(x) = 0$ for all $x \in A^c$.

   Hint. Use the previous exercise.

15. Let A and B be sets in a metric space X and let Y be a second metric space. Suppose that $f : A \cup B \to Y$ is a mapping, such that $f|_A$ and $f|_B$ are continuous. Suppose, further, that $A \setminus B$ is open and closed relative to $A \triangle B$ (the symmetric set-difference). Show that f is continuous.

16. Let X be a metric space. The *support* of a function $f : X \to \mathbb{R}$ is the set

$$\operatorname{supp}(f) := \overline{\{x \in X : f(x) \neq 0\}}.$$

   Show that if f is continuous then $f(x) = 0$ for all $x \in \partial\big(\operatorname{supp}(f)\big)$.

17. Let X and Y be metric spaces and let $f : X \to Y$. Recall the function

$$\omega_f : X \to \mathbb{R} \cup \{\infty\},$$

the point oscillation of f, defined in this section (2.3.3 "Limits of functions"). Show that for every λ (both real λ and ∞) the set

$$\{x \in X : \omega_f(x) < \lambda\}$$

is open.

Note. A function $g : X \rightarrow \mathbb{R}$ is said to be *upper semi-continuous* if the set $g^{-1}(]-\infty, \lambda[)$ is open for every λ. The result here is that ω_f is upper semi-continuous.

18. Let X and Y be metric spaces and let $f : X \rightarrow Y$. Show that the set of all points $x \in X$ at which f is discontinuous is a countable union of closed sets.

   Hint. Use exercise 17, and recall proposition 2.26.

The least subset of $\mathcal{P}(X)$, that includes the open sets and is closed under the operations of complementation and countable union, is called the σ-algebra of Borel sets of X and its elements are called Borel sets. It plays an important role in measure and integration theory. We shall encounter σ-algebras again in section 2.6.

It appears from exercise 18 that the set of discontinuities of a function $f : X \rightarrow \mathbb{R}$ is a Borel set, and one of a rather simple kind. A countable union of closed sets is sometimes termed an F_σ-set. The complementary notion, a countable intersection of open sets, is called a G_δ-set. The mysterious terminology derives from German, σ for 'Summe' and δ for 'Durchsnitt', and can be found in Hausdorff's famous book on set theory.

2.4   Continuity of Linear Mappings

A normed space X over $\mathbb{R}$ has two structures that are bound together:

a) A linear structure, comprising two algebraic operations that satisfy the axioms of a vector space:

   i) Addition of vectors

   $$(x, y) \mapsto x + y : X \times X \rightarrow X$$

   That is, X is an abelian group with neutral element 0.
   ii) Scalar multiplication of a number and a vector

   $$(\lambda, x) \mapsto \lambda x : \mathbb{R} \times X \rightarrow X.$$

b) A metric space structure. It is equipped with a norm $\|x\|$ that satisfies certain axioms (binding it to the linear structure). This gives rise to the metric $d(x, y) = \|x - y\|$.

It is assumed that the reader is familiar with the basic properties of vector spaces and of linear mappings. The notion of a norm and its accompanying metric was introduced in chapter 1.

It is the case that addition and scalar multiplication are continuous operations. This is most easily checked by using the criterion for continuity based on sequences (proposition 2.29). If $x_n, y_n \in X$ ($n \in \mathbb{N}_+$) and $(x_n, y_n) \to (x, y)$ in the product space $X \times X$ then

$$\|(x_n + y_n) - (x + y)\| \leq \|x_n - x\| + \|y_n - y\| \to 0$$

That is, $x_n + y_n \to x + y$, so that addition is a continuous operation.

Again, if $\lambda_n \in \mathbb{R}$, $x_n \in X$ ($n \in \mathbb{N}_+$) and $(\lambda_n, x_n) \to (\lambda, x)$ in the product space $\mathbb{R} \times X$ then

$$\|\lambda_n x_n - \lambda x\| \leq \|(\lambda_n - \lambda)x_n\| + \|\lambda(x_n - x)\|$$
$$= |\lambda_n - \lambda|\|x_n\| + |\lambda|\|x_n - x\|.$$

The premises imply that $|\lambda_n - \lambda| \to 0$ and $\|x_n - x\| \to 0$, from which we obtain $\|x_n\| \to \|x\|$ (the function $x \mapsto \|x\|$ is continuous; it is the distance from x to 0), so that $\lambda_n x_n \to \lambda x$. This shows that scalar multiplication is a continuous operation.

2.4.1 Continuity Criterion

In a normed space X the distance between vectors x and y is the length of the vector $x - y$. The metric is therefore invariant under translation. For all x, y and a in X, we have

$$d(x, y) = d(x - a, y - a).$$

We define the translation of a set $C \subset X$ by the vector a as

$$C + a := \{x + a : x \in C\}.$$

For the ball of radius r we have

$$B_r(x) = B_r(x - a) + a.$$

Let X and Y be normed spaces. The reader should verify that, by the remarks of the previous paragraph, a linear mapping $T : X \to Y$ is continuous if and only if it is continuous at the point 0. This gives rise to a simple criterion for continuity of T. In what follows we do not distinguish notationally between the norm in X and the norm in Y, assuming that the context makes it clear which is intended.

Proposition 2.32 *A linear mapping* $T : X \to Y$ *is continuous if and only if there exists* $K > 0$, *such that* $\|Tx\| \le K\|x\|$ *for all* $x \in X$.

Proof The condition is clearly sufficient since it implies continuity at 0, and hence everywhere. Now suppose that T is continuous. There exists $\delta > 0$, such that $\|Tx\| < 1$ if $\|x\| < 2\delta$. If $x \ne 0$ we have

$$\left\| T\left(\frac{\delta x}{\|x\|} \right) \right\| < 1 \Rightarrow \|Tx\| < \delta^{-1}\|x\|$$

and we can therefore set $K = \delta^{-1}$. $\square$

The conclusion of the preceding proposition can be described as follows: a linear mapping from one normed space to another is continuous if and only if it is bounded on the unit ball, or more generally on bounded sets (exercise 2 below). For this reason a continuous linear mapping between normed spaces is often called, in the context of functional analysis, a bounded linear mapping, though of course it is not bounded over all of X.

The condition also shows that if a linear mapping is continuous it maps Cauchy sequences to Cauchy sequences. It is, in fact, uniformly continuous, a concept, probably familiar to the reader from analysis, but which will be taken up later.

Of great interest in the study of normed spaces are real valued linear mappings. Historically these were called linear functionals, especially when the space in question was a space of functions, and the term probably explains the appellation 'functional analysis' (in the context of which we may also have *non-linear functionals*). The reader may recall that in linear algebra one studies vector spaces over a field F. We may regard F as a one-dimensional vector space over F and a linear mapping from a vector space X (over F) to F is called a linear functional on X. The criterion for continuity, proposition 2.32, applies of course in the case of a linear functional Λ on a normed space X over $\mathbb{R}$.

Continuity of linear mappings is automatic in one important case:

Proposition 2.33 *Let X be a normed space. Then all linear mappings* $T : \mathbb{R}^n \to X$, *where* $\mathbb{R}^n$ *has the norm* $\|a\|_1$, *are continuous.*

Proof Let $(e_1, \ldots, e_n)$ be the standard basis of $\mathbb{R}^n$ and let $u_k = Te_k$, for $k = 1, \ldots, n$. Then $Ta = a_1 u_1 + \cdots + a_n u_n$. Therefore

$$\|Ta\| = \|a_1 u_1 + \cdots + a_n u_n\| \le \sum_{k=1}^{n} |a_k| \|u_k\| \le K\|a\|_1$$

where $K = \max_{1 \le i \le n} \|u_i\|$. We conclude that T is continuous. $\square$

Later we shall see (section 4.4) that this holds with any norm on $\mathbb{R}^n$, but this preliminary result is the key to further progress.

2.4.2   Operator Norms

Let X and Y be normed vector spaces over $\mathbb{R}$. The sum of two continuous linear mappings from X to Y is continuous, and if $T : X \to Y$ is linear and continuous, and if $\lambda \in \mathbb{R}$, then λT is linear and continuous.

Exercise Verify these claims.

Thus the set of all linear and continuous mappings from X to Y is itself a vector space, which we denote by $L(X, Y)$. It can be equipped with a natural norm, the operator norm.

As we have seen, the necessary and sufficient condition that a linear mapping $T : X \to Y$ is continuous is that there exists a constant K such that

$$\|Tx\| \le K\|x\|$$

for all $x \in X$. We therefore define

$$\|T\| = \sup_{x \neq 0} \frac{\|Tx\|}{\|x\|}.$$

This immediately implies the ubiquitously used inequality

$$\|Tx\| \le \|T\|\,\|x\|, \quad (x \in X), \tag{2.1}$$

and the fact, which the reader should verify, that $\|T\|$ is the lowest constant K for which $\|Tx\| \le K\|x\|$ holds for all $x \in X$. Hence, whenever we find a constant K, such that $\|Tx\| \le K\|x\|$ for all vectors x, we can deduce that $\|T\| \le K$. This trivial step is worth pointing out because it is used so frequently.

Some alternative formulations of the same operator norm, taking suprema over vectors in various parts of the open or closed unit ball centred at 0, are listed in the next proposition.

Proposition 2.34

$$\|T\| = \sup_{\|x\| \le 1} \|Tx\| = \sup_{\|x\| < 1} \|Tx\| = \sup_{\|x\| = 1} \|Tx\|.$$

Proof Let A, B and C be the second, third and fourth terms in the above equalities. The definition of $\|T\|$ virtually says that $\|T\| = C$. By (2.1) we have that $A \le \|T\|$, $B \le \|T\|$ and since, obviously, $B \le A$ it is enough to show that $B \ge \|T\|$.

Let $x \neq 0$ and let $0 < \alpha < 1$. By the definition of B we have

$$\|Tx\| = \alpha^{-1}\|x\|\,\|T(\alpha x/\|x\|)\| \le \alpha^{-1} B\|x\|.$$

Letting $\alpha \to 1$ we find

$$\|Tx\| \leq B\|x\|.$$

Since this holds for all $x \neq 0$ we have $B \geq \|T\|$. □

Proposition 2.35 *The operator norm* $\|T\|$ *is a norm on the space* $L(X, Y)$.

Proof Let $\lambda \in \mathbb{R}$. We have

$$\sup_{\|x\|=1} \|(\lambda T)(x)\| = \sup_{\|x\|=1} |\lambda|\|Tx\| = |\lambda| \sup_{\|x\|=1} \|Tx\|.$$

That is, $\|\lambda T\| = |\lambda|\,\|T\|$.

Let S and T be elements of $L(X, Y)$. For all vectors $x \in X$ we have

$$\|(S + T)(x)\| = \|Sx + Tx\| \leq \|Sx\| + \|Tx\|$$

$$\leq \|S\|\,\|x\| + \|T\|\,\|x\| \leq (\|S\| + \|T\|)\|x\|.$$

We deduce that $\|S + T\| \leq \|S\| + \|T\|$. □

Since linear mappings from $\mathbb{R}^m$ to $\mathbb{R}^n$ can be represented as $n \times m$ matrices (using, say, the standard bases in the coordinate spaces $\mathbb{R}^m$ and $\mathbb{R}^n$), the operator norm produces a norm on the vector space of $n \times m$ matrices. Of course, the actual outcome depends on what norms are in use in the coordinate spaces. If the Euclidean norms are in use, then it can be shown that the outcome for the $n \times m$ matrix A is the square root of the highest eigenvalue of the square symmetric matrix A^*A. This is a bit challenging to calculate.

Assume next that we have three normed spaces X, Y and Z. Now if $S \in L(X, Y)$ and $T \in L(Y, Z)$ we can form the composition $T \circ S \in L(X, Z)$. This is usually written without the composition symbol as TS and regarded as a product of some kind. In the case of coordinate spaces $X = \mathbb{R}^\ell$, $Y = \mathbb{R}^m$, $Z = \mathbb{R}^n$ it corresponds to the usual multiplication of matrices.

Exercise Show that $\|TS\| \leq \|T\|\,\|S\|$, the norms being the operator norms in the three corresponding spaces of operators.

The case when all three spaces are the same is important. The normed space $L(X, X)$ is then a so-called normed algebra, in which the (usually non-commutative) product of two elements ST, for $S, T \in L(X, X)$, is actually the composition of mappings, as defined above.

An outstanding problem that arises is the following: does the space $L(X, Y)$ necessarily contain any linear mapping apart from 0? Clearly, in special cases we can construct such mappings; in particular if X and Y are finite-dimensional, then matrices can be used to produce a lot of linear mappings. But in the case that X is an arbitrary infinite-dimensional normed space the answer is not so clear. In the case that $Y = \mathbb{R}$ then the space $L(X, Y)$ is called the normed dual of X. It is denoted by X^*. A fundamental theorem of functional analysis, the Hahn-Banach theorem, is needed to guarantee that the normed dual of a non-trivial but arbitrary normed vector space is not just $\{0\}$. It requires the axiom of choice in the form of Zorn's lemma.

2.4.3 Exercises

1. Let X and Y be vector spaces over $\mathbb{R}$. A mapping $f : X \to Y$ is called *affine* if it satisfies

$$f\big((1-t)u + tv\big) = (1-t)f(u) + tf(v), \quad (u, v \in X, \quad 0 \le t \le 1).$$

   a) Show that f is affine if and only if the mapping $g(x) := f(x) - f(0)$ is linear.

   b) Let X and Y be normed spaces. Show that a continuous mapping $f : X \to Y$, that satisfies

$$f\big(\tfrac{1}{2}(u + v)\big) = \tfrac{1}{2}f(u) + \tfrac{1}{2}f(v), \quad (u, v \in X)$$

   is affine.
   Hint. Show first that the equation

$$f\big((1-t)u + tv\big) = (1-t)f(u) + tf(v)$$

   holds for all t in $[0, 1]$ of the form $m/2^n$, with $m, n \in \mathbb{N}_+$.

2. If X and Y are normed spaces and $T : X \to Y$ a linear mapping, show that T is continuous if and only if it maps bounded sets to bounded sets.

3. Show that the mapping $\pi_n : \ell^p \to \mathbb{R}$, that is given by $\pi_n(a) = a_n$, (the nth coordinate of the sequence a), is a continuous linear functional. It is assumed that $1 \le p \le \infty$.

4. Let $1 < p < \infty$ and let q satisfy $1/p + 1/q = 1$. Let $a = (a_k)_{k=1}^{\infty} \in \ell^q$. Show that the mapping $\lambda_a : \ell^p \to \mathbb{R}$, given by

$$\lambda_a(x) = \sum_{k=1}^{\infty} a_k x_k, \quad (x = (x_k)_{k=1}^{\infty} \in \ell^p)$$

   is a well-defined continuous linear functional on ℓ^p. Well-defined means here that the series $\sum_{k=1}^{\infty} a_k x_k$ is convergent.

Note. A famous theorem of functional analysis says that all continuous linear functionals on ℓ^p are of this form. This is the topic of a future exercise (3.3 exercise 7).

5. Let $a \in [0, 1]$. Show that the function $\lambda_a : C[0, 1] \to \mathbb{R}$, defined by $\lambda_a(f) = f(a)$, is a continuous linear functional.

6. Show that the function $T : C[a, b] \to \mathbb{R}$, defined by $T(f) = \int_a^b f$, is a continuous linear functional. This implies that integral and limit of a function sequence commute if the limit is attained uniformly.

7. Define some continuous linear functionals on $C[a, b]$ that are not just linear combinations of the functionals treated in the previous two exercises.

8. In this exercise we construct a normed space X and a discontinuous linear functional $\Lambda : X \to \mathbb{R}$. Let $E = C[0, 1]$ with its usual norm $\|f\| = \sup_{0 \le x \le 1} |f(x)|$. Let X be the vector space of all $f \in E$, such that f is differentiable at the point $x = \frac{1}{2}$. We equip X with the same norm as E. Define the linear functional $\Lambda : X \to \mathbb{R}$ by $\Lambda(f) = f'(\frac{1}{2})$.

a) Let $\phi : \mathbb{R} \to \mathbb{R}$ be a C^1 function, such that $\phi'(x) > 0$ for all x, $\lim_{x \to -\infty} \phi(x) = 0$ and $\lim_{x \to \infty} \phi(x) = 1$. Let $f_n \in X$ be the function

$$f_n(x) = \phi\big(\tfrac{1}{2} + n(x - \tfrac{1}{2})\big), \quad (0 \le x \le 1).$$

Show that $\|f_n\| \le 1$ but $\Lambda(f_n) \to \infty$.

b) Deduce from (a) that Λ is discontinuous.

Hint. It may help to draw some pictures, with the example $\phi(x) = \frac{1}{2} + \frac{1}{\pi} \arctan x$, to see what happens as n increases.

Note. Every infinite-dimensional normed space admits a discontinuous linear functional. This is because every vector space has a basis, a result that can be proved using Zorn's lemma.

9. Let X be a normed space and let E be a vector subspace of X. Show that the closure $\overline{E}$ is also a vector subspace of E.

10. Let X be a normed space and let E be a hyperplane (a linear subspace of codimension 1). Show that either E is closed or $\overline{E} = X$ (the latter situation is described by saying that E is *dense* in X).

11. Let X be a normed space over $\mathbb{R}$ and let $f : X \to \mathbb{R}$ be a continuous linear functional. Let K be the kernel of f, that is $K = f^{-1}(0)$ (also called the nullspace of f). Let $u \in X$ be such that $f(u) = 1$. Show that

$$\|f\| = \frac{1}{d(u, K)}.$$

12. Let X be a normed space and let Λ be a linear functonal on E. Show that Λ is continuous if and only if its kernel $K := \Lambda^{-1}(0)$ is closed. Use this to give an example of a hyperplane that is not closed.

13. The definition of operator norm as $\sup_{\|x\|=1} \|Tx\|$ raises the question as to whether the supremum is attained. Is it actually the maximum of $\|Tx\|$ on the unit ball? It is simple to give an example when the supremum is attained, using one-dimensional spaces. Here is an example where the supremum is not attained.

The space of all real sequences $(a_n)_{n=1}^{\infty}$, such that $\lim_{n\to\infty} a_n = 0$, is a linear subspace of ℓ^{∞}. Banach denoted it by c_0. It has the same norm as ℓ^{∞}, namely $\sup_{1 \le n < \infty} |a_n|$. Define the linear functional

$$f(a) = \sum_{n=1}^{\infty} 2^{-n} a_n, \quad (a = (a_n)_{n=1}^{\infty} \in c_0).$$

Show that $\|f\| = 1$, but that $|f(a)| < 1$ for all a with $\|a\| = 1$.

14. Let E be a normed space and let V be a closed linear subspace of E. Show a norm can be defined on the quotient space E/V by

$$\|V + x\| = d(x, V).$$

Don't forget to show that this is well defined (that $d(x, V)$ depends only on $V + x$). What can go wrong if V is not closed?

2.5 Homeomorphisms and Topological Properties

Let X and Y be metric spaces.

Definition A mapping $f : X \to Y$ is called a *homeomorphism* (from X to Y) if it is bijective and both f and f^{-1} are continuous.

A bijective mapping f from X to Y sets up a one-to-one correspondence between the points of X and the points of Y. It also commutes with union and intersection, so that $f(\cup A_i) = \cup f(A_i)$ and $f(\cap A_i) = \cap f(A_i)$, and similarly for f^{-1}. In addition to this a homeomorphism sets up a one-to-one correspondence between the open sets of X and the open sets of Y; thus $f(U)$ is open if and only if U is open. Therefore also, $f(U)$ is closed if and only if U is closed. If there exists a homeomorphism from X to Y we say that X and Y are *homeomorphic spaces*.

A common situation is when $f : X \to Y$ is injective but not surjective, and it is a homeomorphism from X to its range $f(X)$. We then say that f is a *topological embedding* of X into Y. It is important to understand that the inverse $f^{-1} : f(X) \to X$ has to be continuous, given that $f(X)$ is viewed as a subspace of Y. This does not follow automatically from the continuity of f.

If $f : X \to Y$ is a homeomorphism then a sequence $(a_n)_{n=1}^{\infty}$ in X is convergent if and only if the corresponding sequence $(f(a_n))_{n=1}^{\infty}$ in Y is convergent. A homeomorphism sets up a one-to-one correspondence between convergent sequences in X and convergent sequences in Y.

Consider the example

$$f :]0, 1[\to \mathbb{R}, \quad f(x) = \tan\left(\pi(x - \tfrac{1}{2})\right).$$

This defines a homeomorphism from $]0, 1[$ to $\mathbb{R}$. Its inverse is the function

$$g : \mathbb{R} \to]0, 1[, \quad g(x) = \tfrac{1}{2} + \tfrac{1}{\pi} \arctan x.$$

The continuity of f and g is proved in beginning courses in analysis.

We observe that f maps the sequence $1/n$, $(n = 1, 2, \ldots)$, to the sequence $\tan(\pi(\tfrac{1}{n} - \tfrac{1}{2}))$, $(n = 1, 2, \ldots)$. This might appear to contradict the statement that a homeomorphism sets up a correspondence between convergent sequences, since $\lim_{n \to \infty} f(1/n) = -\infty$ whilst $\lim_{n \to \infty} 1/n = 0$. The point is that we must view $]0, 1[$ as a metric space in its own right, and, when viewed as such, the sequence $1/n$, $(n = 1, 2, \ldots)$, is divergent.

If X and Y are homeomorphic spaces, then any property of X that can be expressed in terms of its open sets together with set-theoretical sentences, is also possessed by Y. Such properties are called *topological properties* and are often the most significant that a space may possess. We shall provide further clarification of these properties in section 2.6.

If a space X has a topological property that Y does not have, then X and Y cannot be homeomorphic; there exists no homeomorphism $f : X \to Y$. Because one often wants to know whether two spaces are homeomorphic, or, more particularly, one wishes to disprove that they are so, it is important to have an extensive list of topological properties. So let's list some topological properties that we can define with the material at hand. In some cases we also mention where, later in this text, the corresponding property will be studied. The first examples are rather dull.

a) Any set theoretical property, for example, X has cardinality **c**.
b) Every subset of X is open. This property is described by saying that X is a *discrete space*. Note that this is the same as saying that every point of X is isolated (2.2 exercise 2). It can be achieved by the metric $d(x, y) = 1$ if $x \neq y$.
c) The only open sets are $\emptyset$ and X. This property is described by saying that X is an *indiscrete space*. It is not achievable by a metric, except when X has cardinality 1.
d) X includes a countable set A such that $\overline{A} = X$. This is an important property, called *separability*. It is clearly topological since anything that can be expressed by closed sets can also be expressed by open sets. We shall study this property in chapter 5.
e) Every sequence in X has a convergent subsequence. This is a topological property since convergence of a sequence can be expressed by using open sets. This is an immensely important property, called *sequential compactness*. We shall study this property in chapter 4.

f) Every continuous function $f : X \to \mathbb{R}$ attains a maximum. This is studied in chapter 4.

g) Every continuous mapping $f : X \to X$ has a fixed point (that is, a solution to $f(x) = x$ exists). Fixed points will be studied in chapter 6.

The reader may recall the Bolzano-Weierstrass theorem from fundamental analysis, to the effect that every bounded sequence of real numbers has a convergent subsequence. In particular every sequence in the interval $[0, 1]$ has a convergent subsequence, and since the set $[0, 1]$ is closed in $\mathbb{R}$, the limit of the convergent subsequence is again in $[0, 1]$. So the *metric space* $[0, 1]$, viewed as a metric space in its own right, is sequentially compact (property (e) in the above list).

We can use this to answer the question: does a homeomorphism

$$f : [0, 1] \to]0, 1[$$

exist? The answer is no; for whilst $[0, 1]$ is a sequentially compact space, $]0, 1[$ is not. It is a test of the reader's understanding to supply an example of a sequence in $]0, 1[$ that has no convergent subsequence. Note that the space $[0, 1]$ also has properties (f) and (g), neither of which is shared by $]0, 1[$. Of course one can answer this question by observing that no continuous surjective mapping from $[0, 1]$ to $]0, 1[$ can exist, since such a mapping could not attain a maximum, contrary to a basic theorem of analysis. However, the proof by unshared topological properties illustrates a valuable method that can be applied in more complicated problems.

Any property that can be predicated of metric spaces, and not just topological properties, can be extended in meaning so that they can be predicated of sets in metric spaces. This a common occurrence. Suppose that ψ is a predicate of metric spaces. Let (X, d) be a metric space and let $A \subset X$. We extend the meaning so that the sentence 'the set A is ψ (in X)' means 'the metric space (A, d) is ψ'. Thus we can say 'A is a complete set (in X)' meaning 'the metric space (A, d) is complete', or 'A is a discrete set (in X)' meaning that the metric space (A, d) is discrete, and so on.

2.5.1 Equivalent Metrics

Some properties of metric spaces are not topological. An example is the property that X is bounded. We saw that $]0, 1[$ is homeomorphic to $\mathbb{R}$, but one is bounded and the other not. Again $\mathbb{R}$ is complete whereas $]0, 1[$ is not. Completeness is not a topological property, but still it is a very important property.

Definition Two metrics d and d' on the same set X are said to be equivalent if the same sets are open in (X, d) as are open in (X, d').

In other words, d and d' are equivalent when the identity mapping from (X, d) to (X, d') is a homeomorphism. In particular the same sequences are convergent for the metric d as for the metric d', and have the same limit.

The following proposition details a useful situation when two metrics can be proved equivalent. The condition given is not necessary for equivalence.

Proposition 2.36 *Let d and d' be metrics on X. Assume that there exist $K > 0$ and $K' > 0$, such that*

$$K'd'(x, y) \leq d(x, y) \leq Kd'(x, y)$$

for all x and y in X. Then the metrics d and d' are equivalent.

Proof It should be obvious that the same sequences are convergent and have the same limits, whether d or d' is used. Hence the same sets are open (by proposition 2.12). □

2.5.2 Exercises

1. Let X and Y be metric spaces and let $f : X \to Y$. The graph of f is the set

$$G_f := \{(x, y) \in X \times Y : y = f(x)\}.$$

   a) Show that if f is continuous then G_f is closed in $X \times Y$, but that the converse is not, in general, true.
   b) Show that f is continuous if and only if the mapping that assigns $(x, f(x))$ to $x \in X$ is a homeomorphism of X to G_f.

2. If X and Y are metric spaces and $f : X \to Y$ is continuous and bijective, it is not necessarily the case that f^{-1} is continuous. Consider the example $f :]0, 2\pi[\to \mathbb{R}^2$ given by

$$f(t) = (\sin t, \sin 2t), \quad (0 < t < 2\pi).$$

   Show that f is continuous, injective, but that $f^{-1} : C \to]0, 2\pi[$, where C is the range of f, is not continuous.
3. Given that $p \in [1, \infty[\cup \{\infty\}$, show that the metrics $\|x - y\|_p$ defined on $\mathbb{R}^n$ are all equivalent to one another.
4. Suppose that two metrics d and d' satisfy the conditions of proposition 2.36. Show that a sequence is a Cauchy sequence with respect to d if and only if it is a Cauchy sequence with respect to d'. In short, d and d' have the same Cauchy sequences.
5. Let d and d' be metrics on a set X. Denote the open ball with centre a and radius r with respect to d by $B_r(a)$, and with respect to d' by $B'_r(a)$. Prove that d and

d' are equivalent metrics if and only if the following conditions are satisfied for every $a \in X$: for every $r > 0$ there exists $s > 0$, such that $B_s'(a) \subset B_r(a)$; and for every $r > 0$ there exists $s > 0$, such that $B_s(a) \subset B_r'(a)$.

6. Let (X, d) be a metric space and let f be a function that satisfies the conditions of 1.1 exercise 5 with the additional condition that f is continuous at 0. Show that the metric $d'(a, b) := f(d(a, b))$ is equivalent to the metric d.
   Hint. There is an interval $[0, \lambda[$ in which f is strictly increasing and so f^{-1} is continuous in $[0, f(\lambda)[$.

7. Show that given a metric space there exists a bounded metric space homeomorphic to it.

8. The n-sphere S^n is the subspace of $\mathbb{R}^{n+1}$ with the Euclidean norm consisting of all unit vectors. Let P be the point $(0, \ldots, 0, 1) \in S^n$ (the North Pole). We shall describe a homeomorphism from $S^n \setminus \{P\}$ on to $\mathbb{R}^n$ called stereographic projection.
   Think of points in S^n as pairs (x, s) with $x \in \mathbb{R}^n$, $s \in \mathbb{R}$ and $|x|^2 + s^2 = 1$. Define $F : \mathbb{R}^n \to S^n \setminus \{P\}$ by

$$F(x) = \left(\frac{2x}{|x|^2 + 1}, \frac{|x|^2 - 1}{|x|^2 + 1} \right).$$

Show that F is a homeomorphism.

Hint. Compute its inverse.

2.6   Topologies and σ-Algebras

As we have seen, a *topological property* of a metric space is one that can be expressed in terms of its open sets, together with set-theoretical sentences. Thus the property of (X, d), that every sequence in X is convergent, is topological, since the property that a sequence converges to a point p can be expressed in terms of open sets containing p, whereas the property that (X, d) is bounded is not. One can argue that what is meant by 'expressed' in the notion of topological property is unclear.

We often study situations where the actual metric does not seem to matter so much as the open sets themselves. We saw that $\mathbb{R}^n$ admits a wide range of metrics $d_p(x, y) = \|x - y\|_p$, for any p in the range $1 \leq p \leq \infty$, but the same sets were open for whichever of these metrics was used. Again, the product of metric spaces (discussed in detail later) admits a variety of metrics, that all define the same open sets. If the same sets are open with respect to two different metrics, then the same mappings will be continuous.

Because of these considerations the idea arose to make the open sets the primary objects of interest instead of the metric and to define their basic properties as axioms. Recall that for a set X the set $\mathcal{P}(X)$ is the power set of X, the set of all subsets of X.

Definition Let X be a set. A subset $\mathcal{T} \subset \mathcal{P}(X)$ is called a topology on X if the following hold:

1) $\emptyset \in \mathcal{T}$ and $X \in \mathcal{T}$.
2) If $(A_i)_{i \in I}$ is a family in $\mathcal{T}$ then $\bigcup_{i \in I} A_i \in \mathcal{T}$.
3) If $(A_i)_{i=1}^{n}$ is a finite family in $\mathcal{T}$ then $\bigcap_{i=1}^{n} A_i \in \mathcal{T}$.

Definition A topological space is a set X together with a topology $\mathcal{T}$ on X. More formally, it is an ordered pair $(X, \mathcal{T})$, where $\mathcal{T}$ is a topology on X.

If X is equipped with a topology $\mathcal{T}$, the sets of $\mathcal{T}$ are called the open sets of the topology, or, more colloquially, the sets open in X. Properties 2 and 3 reproduce the conclusions of propositions 2.7 and 2.8, but are stated as axioms.

The concepts of closed sets, neighbourhoods, continuous mappings, convergence of sequences, and so on, are then developed on the basis of the open sets alone. Some indications of how this is done are given below. Now we can say that topological properties are precisely those properties that a topological space may or may not possess, that is to say, properties that can be expressed for topological spaces.

A metric space (X, d) becomes a topological space when X is equipped with the topology comprising all open sets defined by the metric. However, a given topology on X is not necessarily derivable from a metric on X. It is an important fact that the study of topological spaces is more general than that of metric spaces. In fact let X be a non-empty set. There are two extreme topologies that we can define on X:

a) The topology $\mathcal{T} = \mathcal{P}(X)$. All sets are open. This is called the *discrete topology*. A space equipped with the discrete topology is called a discrete space.
b) The topology $\mathcal{T} = \{\emptyset, X\}$. The only open sets are $\emptyset$ and X. This called the *indiscrete topology*. A space equipped with the indiscrete topology is called an indiscrete space.

Exercise Show that the indiscrete topology cannot arise from a metric if X has at least two points. Show, on the other hand, that the discrete topology can be obtained from the metric given by $d(x, y) = 1$ for $x \neq y$.

Some of the basic definitions used in topological spaces are as follows:

1. A set $A \subset X$ is said to be closed if A^c is open.
2. A set W is said to be a neighbourhood of a point p if there exists an open set U such that $p \in U \subset W$.
3. Given two topological spaces X and Y, each equipped with its topology, a mapping $f : X \to Y$ is said to be continuous at a point $p \in X$ if the following condition is satisfied: for every neighbourhood W of $f(p)$ in Y, the inverse image $f^{-1}(W)$ is a neighbourhood of p in X.
4. A sequence $(x_n)_{n=1}^{\infty}$ in X is said to converge to a point $p \in X$ if the following condition is satisfied: for every neighbourhood W of p there exists a natural number N, such that $x_n \in W$ for all $n \geq N$.
5. Let $A \subset X$. A point p is said to be a limit point of A if the following condition is satisfied: every neighbourhood of p contains a point of A distinct from p.

The notion of topological space is very general, and gives rise to a rich theory. Various extra conditions, necessarily topological, are employed that tend to exclude the more "pathological" cases; the stronger these are, the richer is the theory of the corresponding spaces. Some of these, together with their usual appellations, are:

a) A topological space is said to be *Hausdorff* (that's an adjective, not a noun!) if it satisfies the following condition: for any two points x and y in X, such that $x \neq y$, there exist open sets U and V, such that $x \in U$, $y \in V$ and $U \cap V = \emptyset$.
b) A topological space is said to be *normal* if it satisfies the following condition: if F and G are closed and disjoint, then there exist open sets U and V, such that $F \subset U, G \subset V$ and $U \cap V = \emptyset$.

Exercise Show that metric spaces are both Hausdorff and normal.

Hint. Both properties have appeared already, without name, in this chapter.

One of the important applications of topological spaces (other than metric spaces) to analysis is in the study of so-called weak topologies on normed spaces, and their applications to extremal problems, such as arise in the calculus of variations. For a brief introduction to weak topologies, see exercise 12 in this section.

A topology on a set X is a subset of $\mathcal{P}(X)$ that satisfies certain axioms. It is used to impose a structure on X, that of topological space. Another type of structure, also a subset of $\mathcal{P}(X)$ that satisfies certain axioms, plays a fundamental role in modern theories of integration and in probability theory. We refer to the notion of σ-algebra on the set X.

Definition A subset $\mathcal{M}$ of $\mathcal{P}(X)$ is called a σ-algebra on X if it has the following properties:

1) $\emptyset \in \mathcal{M}$.
2) If $A \in \mathcal{M}$ then $A^c \in \mathcal{M}$.
3) If $(A_k)_{k=1}^{\infty}$ is a sequence of elements of $\mathcal{M}$ then $\bigcup_{k=1}^{\infty} A_k \in \mathcal{M}$.

A set X equipped with a σ-algebra, more precisely an ordered pair $(X, \mathcal{M})$, is called a measurable space.[3] The most important thing that a σ-algebra can do is be the domain of a measure.

---

[3] The idea that a set with a structure is an ordered pair is quite conventional and often convenient. Nevertheless it may be objected that it is superfluous. In the case of a measurable space $(X, \mathcal{M})$ the set X is the maximum element of $\mathcal{M}$ and does not need to be mentioned explicitly. A similar remark applies to viewing a topological space as an ordered pair $(X, \mathcal{T})$ where $\mathcal{T}$ is a topology on X.

Definition Let $(X, \mathcal{M})$ be a measurable space. A function $m : \mathcal{M} \to [0, \infty]$ is called a measure if it has the following property:

If $(A_k)_{k=1}^{\infty}$ is a pair-wise disjoint sequence of elements of $\mathcal{M}$ (that is $A_i \cap A_j = \emptyset$ if $i \neq j$) then

$$m\left(\bigcup_{k=1}^{\infty} A_k\right) = \sum_{k=1}^{\infty} m(A_k).$$

This defining property of a measure is called *σ-additivity* and a measure is said to be *σ-additive*. The pervasive use of the letter 'σ' probably alludes to 'sum', specifically to infinite sum.

Note that the codomain of a measure is given as $[0, \infty]$. This denotes the non-negative reals together with the element ∞. Addition in $[0, \infty]$ is defined in the usual way for two finite elements, whilst we set $a + \infty = \infty$. The sum of the infinite series on the right, if divergent, is taken to be ∞.

A measure space is a triple $(X, \mathcal{M}, m)$ where $\mathcal{M}$ is a σ-algebra on X and m a measure with domain $\mathcal{M}$.

Some simple examples of measure spaces are:

a) $X = \mathbb{N}_+$, $\mathcal{M} = \mathcal{P}(\mathbb{N}_+)$, $m(A)$ is the number of elements in A if A is finite, ∞ if A is infinite. This is called counting measure on $\mathbb{N}_+$.
b) X is a set, $a \in X$, $\mathcal{M} = \mathcal{P}(X)$, $m(A) = 1$ if $a \in A$, 0 otherwise. This measure is called a unit mass at the point a.

Constructing a σ-algebra for a set X, and a measure defined on it, can be a difficult task, except for some rather simple examples like these. Historically the construction of Lebesgue measure in $\mathbb{R}$ was the most important challenge. It was sought to extend the notion of length of an interval to a measure. More precisely, to find a σ-algebra $\mathcal{L}$ on $\mathbb{R}$ (a set of subsets of $\mathbb{R}$ with the properties specified above), that would include all intervals $[a, b]$, and a measure m with domain $\mathcal{L}$ that assigned the value $b - a$ to the interval $[a, b]$.

If X is a topological space then there is an exceedingly important σ-algebra on X, the one generated by the set of all open sets. It is the minimum (in the sense of set inclusion) σ-algebra that includes all the open sets. This is called the Borel algebra and its elements are called Borel sets. A measure defined on the Borel sets is called a Borel measure.

2.6.1 Order Topologies

In analysis with one real variable (fundamental analysis) the notions of convergence and continuity do not require one to view $\mathbb{R}$ as a metric space. In fact the statement $\lim_{n \to \infty} a_n = b$ can be viewed as meaning:

For all $\varepsilon > 0$ there exists N, such that $b - \varepsilon < a_n < b + \varepsilon$ for all $n \geq N$.

Similarly the statement that a function $f : \mathbb{R} \to \mathbb{R}$ is continuous at a can be viewed as meaning:

> *For all $\varepsilon > 0$ there exists $\delta > 0$, such that $f(a) - \varepsilon < f(x) < f(a) + \varepsilon$ for all x that satisfy $a - \delta < x < a + \delta$.*

All that is required here is the ordering of $\mathbb{R}$, and these statements actually make sense in any ordered field. An ordered field is not necessarily a metric space, a fact which is perhaps not obvious, but clearly we are not appealing here to any metric to make sense of continuity and convergence.

Actually, positivity, apparently called upon when we say "For all $\varepsilon > 0$", is not really needed. We can define continuity and convergence using intervals alone. In fact any totally ordered set can be equipped with a topology, called the *order topology*. Let $(X, \preceq)$ be a totally ordered set. In imitation of analysis in $\mathbb{R}$, we call the sets

$$]u, v[:= \{x \in X : u \prec x \prec v\}, \quad (u \prec v)$$

$$]u, \infty[:= \{x \in X : u \prec x\}$$

$$]-\infty, v[:= \{x \in X : x \prec v\},$$

together with $\emptyset$ and X itself, open intervals. The formula '$u \prec v$' is short for '$u \preceq v \wedge u \neq v$', since, by convention, the actual order relation, symbolised by '$\preceq$', is taken to be reflexive. The endpoints u and v are arbitrary points of X. The symbols ∞ and $-\infty$ do not denote points of X; they are therefore not endpoints. In fact the sets $]u, \infty[$ and $]-\infty, v[$, having only one endpoint, are often called *rays*.

Now we say that a set $A \subset X$ is *open* if it satisfies the following condition: for every $p \in A$ there exists an open interval I such that $p \in I \subset A$.

Exercise Show that the set of all open sets A is a topology $\mathcal{T}$ on X.

Hint. Show that if I and J are open intervals, and if a point p is in $I \cap J$, then there exists an open interval K such that $p \in K \subset I \cap J$.

The topology $\mathcal{T}$ is called the *order topology* on X. Because it is the order topology on $\mathbb{R}$ that is used to define convergence and continuity in fundamental analysis, one can argue that metric space theory is not really a foundational theory for analysis. It is instead a tool that facilitates to an extraordinary degree the study of multivariate calculus, functional analysis, and all their derivative subjects.

One final point: the notion of boundedness makes sense in a totally ordered set. A subset A is *bounded above* if there exists y such that $x \preceq y$ for all $x \in A$. Such a point y is called an upper bound for A. In a similar way we define the concept of bounded below. The set A is then *bounded* when it is both bounded above and bounded below—no metric required here.

2.6.2 Exercises

1. A set X contains exactly two elements. List all the possible topologies on X.
2. Let $(X, \mathcal{T})$ be a topological space. Show that a set U is open if and only if it is a neighbourhood of each of its points.
3. Let $(X, \mathcal{T}_1)$ and $(Y, \mathcal{T}_2)$ be topological spaces and let $f : X \to Y$. Show that f is continuous at all points in X (in short, f is continuous) if and only if the following condition is satisfied: for all $U \subset Y$, the inverse image $f^{-1}(U)$ is open in X whenever U is open in Y.

   Note. This reproduces proposition 2.23 for general topological spaces.

4. Let X be a totally ordered set. Show that the order topology on X is Hausdorff.
5. For this exercise some knowledge of well-ordered sets is required. Let X be a well-ordered set and equip X with the order topology.

   a) Show that every increasing sequence in X that is bounded above is convergent.

      Hint. Consider the least upper bound of the terms of the sequence.
   b) Show that every bounded sequence has a convergent subsequence.

      Hint. In a totally ordered set, every sequence has a monotonic subsequence. This may be known to the reader for real sequences, in which context it is sometimes used to prove the Bolzano-Weierstrass theorem. The proof for totally ordered sets is the same.

6. Let X and Y be totally ordered sets each equipped with its order topology. Show that a function $f : X \to Y$ is continuous at the point $p \in X$ if and only if the following condition is satisfied: for every open interval J in Y such that $f(p) \in J$ there exists an open interval I in X such that $p \in I$ and $f(I) \subset J$.

   Note. This virtually reproduces the definition of continuity of real analysis for a function $f : \mathbb{R} \to \mathbb{R}$: for every $\varepsilon > 0$ there exists $\delta > 0$ such that

   $$f\big(]p - \delta, p + \delta[\big) \subset \,]f(p) - \varepsilon, \, f(p) + \varepsilon[.$$

7. Let X be a set and let $(\mathcal{T}_i)_{i \in I}$ be a family of topologies on X. Show that their intersection is a topology on X. Hence show that, given a subset $\mathcal{S} \subset \mathcal{P}(X)$, there exists a minimum topology that includes $\mathcal{S}$. It is called the *topology generated by* $\mathcal{S}$.
8. Let X be a set and let $\mathcal{S} \subset \mathcal{P}(X)$. Show that the topology generated by $\mathcal{S}$ (see the previous exercise) consists of the set X together with all sets U that satisfy the following condition: for all $x \in U$ there exist finitely many elements $\{S_1, \ldots, S_n\}$ of $\mathcal{S}$, such that

   $$x \in \bigcap_{k=1}^{n} S_k \subset U. \tag{2.2}$$

9. Closely related to the previous exercise is the notion of *subbase for a topology*.
   Let $(X, \mathcal{T})$ be a topological space. A subbase for the topology $\mathcal{T}$ is a subset
   $\mathcal{S} \subset \mathcal{T}$ with the following property: if U is open and $x \in U$ then there exist
   finitely many elements $\{S_1, \ldots, S_n\}$ of $\mathcal{S}$, such that (2.2) holds.

   a) Show that if $\mathcal{S}$ is a subbase for $\mathcal{T}$ then $\mathcal{S}$ generates $\mathcal{T}$.
   b) Show conversely, that if $\mathcal{S} \subset \mathcal{P}(X)$ and $\mathcal{S}$ generates the topology $\mathcal{T}$, then $\mathcal{S}$
      is a subbase for $\mathcal{T}$ *provided $\mathcal{S}$ covers X.* The latter means that $X = \bigcup \mathcal{S}$.

10. As for topologies, so for σ-algebras. Let X be a set and let $(\mathcal{M}_i)_{i \in I}$ be a family
    of σ-algebras on X. Show that their intersection is a σ-algebra on X. Hence
    show that, given a subset $\mathcal{S} \subset \mathcal{P}(X)$, there exists a minimum σ-algebra that
    includes $\mathcal{S}$. It is called the σ-algebra generated by $\mathcal{S}$.
    Note. We recall that if X is a topological space the σ-algebra generated by the
    open sets is called the Borel algebra of X.

11. Let X be a totally ordered set. Show that the order topology on X is generated
    by the set of all rays $]-\infty, u[$ and $]v, \infty[$, for $u, v \in X$.

12. Let X be a vector space over $\mathbb{R}$ or $\mathbb{C}$ and let M be a vector space consisting of
    linear functionals on X. Note that no norm is imposed on X and that M does
    not necessarily contain all possible linear functionals. For each $x \in X$, $\varepsilon > 0$
    and $f \in M$ we define

    $$U(x, f, \varepsilon) = \{y \in X : |f(y) - f(x)| < \varepsilon\}.$$

    We define a subset $\mathcal{T} \subset \mathcal{P}(X)$ by stipulating that $A \in \mathcal{T}$ if and only if the
    following condition holds: for all $x \in A$ there exist $\varepsilon > 0$ and a finite set f_1,
    $\ldots, f_m$ of elements of M, such that

    $$\bigcap_{k=1}^{m} U(x, f_k, \varepsilon) \subset A.$$

    a) Show that $\mathcal{T}$ is a topology for X, for which the set of all sets of the form
       $U(x, f, \varepsilon)$ is a subbase.
    b) Show that the topology $\mathcal{T}$ is Hausdorff if and only if the following condition
       is satisfied: for all non-zero x in X, there exists $f \in M$ such that $f(x) \neq 0$.
    c) Show that if X is equipped with the topology $\mathcal{T}$ then all the linear functionals
       in M are continuous.
    d) Prove the converse: if f is a linear functional on X, continuous for the
       topology $\mathcal{T}$, then $f \in M$.

       Hint. If f is continuous at 0 there exists $\delta > 0$ and $\{f_1, \ldots, f_m\} \subset M$,
       such that $|f(y)| < 1$ for all y in the set $\bigcap_{k=1}^{m} U(0, f_k, \delta)$. Deduce that if
       $f_k(y) = 0$ for $k = 1, \ldots, m$ then $f(y) = 0$ and therefore, by linear algebra,
       f is a linear combination of the functionals $f_1, \ldots, f_m$.

Note. The topology $\mathcal{T}$ is called the weak topology on X induced by M. Weak topologies are important in functional analysis, and some of its applications, most notably calculus of variations. If X is a normed space and X^* the normed space of continuous linear functionals on X (the normed dual of X, see section 2.4), then the weak topology induced on X by X^* is simply called *the weak topology* on X. It is not derivable from a metric if X is infinite-dimensional. Each element x of X can be viewed as a linear functional on X^*, namely by the prescription $f \mapsto f(x)$. This produces the weak topology on X^* induced by X, usually called *the weak* topology on X^**.

13. For this exercise a knowledge of well-ordered sets and infinite cardinals is required. It can be shown that there exists an uncountable well-ordered set such that the initial segment of every element is countable.[4] Every such set has the same cardinal; it is denoted by $\aleph_1$ and is the lowest uncountable cardinal. The unresolved, and even unresolvable, continuum hypothesis is the assertion that $\aleph_1 = \mathbf{c}$.

In more detail there is an uncountable well-ordered set W such that for all $x \in W$ the set

$$I_W(x) := \{y \in W : y \prec x\}$$

(the initial segment of x) is countable. Adjoin a new element Ω, not in W, and order the set $W \cup \{\Omega\}$ by positing that Ω lies above every element of W.

a) Show that W has no maximum element.

b) Show that if W is equipped with the order topology and $(x_n)_{n=1}^{\infty}$ is a sequence in W, then $(x_n)_{n=1}^{\infty}$ is bounded. Deduce that every sequence in W has a convergent subsequence.

   Hint. What can you say about the union

$$\bigcup_{n=1}^{\infty} I_W(x_n)?$$

   For the second deduction use exercise 5.

c) Show that in the order topology of $W \cup \{\Omega\}$, the element Ω is a limit point of the subset W, but that no sequence in W exists that has the limit Ω.

d) Deduce that the order topology on $W \cup \{\Omega\}$ cannot be derived from a metric.

---

[4] This follows from the well-ordering theorem, that every set can be well-ordered.

2.6.3   *Pointers to Further Study*

$\rightarrow$ General topology.
$\rightarrow$ Measure theory.
$\rightarrow$ Probability theory.
$\rightarrow$ Weak topologies in Banach spaces.

2.7   ($\Diamond$) Mazur-Ulam

Let (X, d_X) and (Y, d_Y) be metric spaces. A mapping $f : X \rightarrow Y$ is called an *isometry* if, for all x and x' in X we have

$$d_Y\big(f(x), f(x')\big) = d_X(x, x').$$

In other words an isometry is a mapping that preserves the metric. An isometry is necessarily injective but not necessarily bijective; throughout this section surjectivity will not be required of an isometry. This seems to be the most common usage, but it is admittedly inconsistent with the way we say that $f : X \rightarrow Y$ is a homeomorphism, which is usually taken to mean that f is surjective.

We say that spaces X and Y are *isometric* (to each other) if there exists a bijective isometry from X to Y; or, as we would usually say, an isometry of X *on to* Y. In a sense, isometric metric spaces are identical as regards all their metric space properties.

Let X and Y be normed vector spaces. Recall that a mapping $f : X \rightarrow Y$ is called affine (2.4 exercise 1) if it satisfies

$$f\big((1 - t)u + tv\big) = (1 - t)f(u) + tf(v), \quad (u, v \in X, \quad 0 \le t \le 1).$$

Obviously an affine mapping satisfies

$$f\big(\tfrac{1}{2}(u + v)\big) = \tfrac{1}{2}f(u) + \tfrac{1}{2}f(v),$$

a property that we could describe, with some risk of misunderstanding, by saying that f is midpoint preserving. There is a partial converse to this which we are going to exploit: a continuous mapping that is midpoint preserving is affine. This was 2.4 exercise 1(b).

Proposition 2.37 (Mazur-Ulam Theorem) *Let X and Y be normed spaces and suppose that $f : X \rightarrow Y$ is a bijective isometry. Then f is affine.*

Proof Since an isometry is continuous, it is enough to show that f is midpoint preserving. This is accomplished in two steps.

Step 1. Let $u \in X$ and $v \in X$. We show that if $g : X \to X$ is an isometry that satisfies $g(u) = u$ and $g(v) = v$ then it also satisfies $g\left(\frac{1}{2}(u + v)\right) = \frac{1}{2}(u + v)$.

Let $z = \frac{1}{2}(u + v)$ and let W be the set of all isometries $g : X \to X$ that fix u and v. Define

$$\lambda = \sup\{\|g(z) - z\| : g \in W\}.$$

We shall show that $\lambda = 0$, thus proving that $g(z) = z$ for all g in W.

Firstly, $\lambda < \infty$. For let $g \in W$. Then

$$\|g(z) - z\| \leq \|g(z) - u\| + \|u - z\|$$
$$= \|g(z) - g(u)\| + \|u - z\| = 2\|u - z\|.$$

Hence

$$\lambda \leq 2\|u - z\| < \infty.$$

Again we let $g \in W$. Define the mapping $\psi : X \to X$ by

$$\psi(x) = 2z - x.$$

Obviously ψ is an isometry that fixes z and interchanges u and v. So the composed mapping $\psi \circ g^{-1} \circ \psi \circ g$ is an isometry that fixes u and v. Therefore

$$2\|g(z) - z\| = \|(2z - g(z)) - g(z)\|$$
$$= \|(\psi \circ g)(z) - g(z)\|$$
$$= \|(g^{-1} \circ \psi \circ g)(z) - z\|$$
$$= \|(g^{-1} \circ \psi \circ g)(z) - \psi^{-1}(z)\|$$
$$= \|(\psi \circ g^{-1} \circ \psi \circ g)(z) - z\|$$

Now $\psi \circ g^{-1} \circ \psi \circ g$ is an element of W. Hence the last term is less than or equal to λ and we conclude that

$$2\|g(z) - z\| \leq \lambda.$$

As g is an arbitrary element of W we see that $2\lambda \leq \lambda$; that is, $\lambda = 0$. This concludes step 1.

Step 2. We show that if $u, v \in X$ then $f\left(\frac{1}{2}(u + v)\right) = \frac{1}{2}(f(u) + f(v))$.

Let $z = \frac{1}{2}(u + v)$ and $z' = \frac{1}{2}(f(u) + f(v))$. Define the mapping $\phi : Y \to Y$ by

$$\phi(y) = 2z' - y.$$

Then ϕ is an isometry that fixes z' and interchanges $f(u)$ and $f(v)$. Recall the isometry ψ from step 1 that fixes z and interchanges u and v. The composed mapping $\psi \circ f^{-1} \circ \phi \circ f$ is an isometry on X that fixes u and v. Hence, by step 1, it fixes z also. That is,

$$(\psi \circ f^{-1} \circ \phi \circ f)(z) = z.$$

Therefore

$$(\phi \circ f)(z) = f(z),$$

that is,

$$2z' - f(z) = f(z),$$

or $f(z) = z'$, as required.   □

2.7.1 Exercises

1. Giving $\mathbb{R}^2$ the norm $\|(x, y)\|_\infty = \max(|x|, |y|)$, find an example of an isometry $f : \mathbb{R} \to \mathbb{R}^2$ that is not affine.

   Note. This does not contradict proposition 2.37 since the required isometry will not be surjective.

2. For this exercise you will need some knowledge of linear algebra. Describe all isometries of $\mathbb{R}^2$ on to itself when it is equipped with:

   a) The Euclidean norm $|u| := \|u\|_2$.
   b) The norm $\|u\|_\infty$.
   c) The norm $\|u\|_1$.

3. Repeat the previous exercise for the space $\mathbb{R}^3$.

4. View the Euclidean sphere S^n as the set of all vectors in $\mathbb{R}^{n+1}$ with Euclidean length 1. It is a subspace of $\mathbb{R}^{n+1}$ with the Euclidean metric. Find all isometries of S^n on to itself.

   Hint. A mapping $f : S^n \to S^n$ can be extended to a mapping $F : \mathbb{R}^{n+1} \to \mathbb{R}^{n+1}$ by setting $F(x) = |x| f(x/|x|)$ for $x \neq 0$ and $F(0) = 0$.

   Note. The intended distance between points u and v in S^n is the chordal distance $|u - v|$. An alternative distance that would be perhaps more natural is the great circle arclength or $\widehat{uv}$. By trigonometry this is given by $\widehat{uv} = 2 \arcsin\left(\frac{1}{2}|u - v|\right)$, so that an isometry for one is an isometry for the other.

5. Let X be a normed space and let u and v be points in X. Define the set

$$B_1 = \left\{x \in X : \|x - u\| = \|x - v\| = \tfrac{1}{2}\|u - v\|\right\},$$

and by induction define the sequence $(B_n)_{n=1}^{\infty}$ of sets by

$$B_{n+1} = \left\{ x \in B_n : \|x - y\| \leq \tfrac{1}{2}\mathrm{diam}(B_n) \text{ for all } y \in B_n \right\}, \quad (n \geq 1)$$

Prove that the intersection $\bigcap_{n=1}^{\infty} B_n$ consists of the single point $\tfrac{1}{2}(u + v)$.

Hint. Show that (i) $\mathrm{diam}(B_{n+1}) \leq \tfrac{1}{2}\mathrm{diam}(B_n)$ and (ii) (by induction)

$$\tfrac{1}{2}(u + v) \in B_n \quad \text{and} \quad x \in B_n \Rightarrow u + v - x \in B_n.$$

Note. The exercise shows that the midpoint of a segment with endpoints u and v may be specified as a function $\mathrm{mid}(u, v)$ of u and v, *building only on the metric of X* (recall that $d(x, y) = \|x - y\|$). Nowhere need the linear structure as such be referenced. It is straightforward to verify that from the midpoint function $\mathrm{mid}(u, v)$, together with a knowledge of the zero-vector and the metric, one can recover the whole linear structure. So the metric and the zero-vector determine the linear structure in a normed vector space.

6. Using the previous exercise one can give another proof of the Mazur-Ulam theorem, one that gives a deeper insight into it.

Let X and Y be normed spaces and let $f : X \to Y$ be a bijective isometry that also maps 0 to 0. Define a new linear structure in X by defining the new sum of x and y to be $f^{-1}(f(x) + f(y))$ and the new product of λ and x to be $f^{-1}(\lambda f(x))$. Observe that the norm of X with the old structure is also a norm with the new structure and derive proposition 2.37 by applying the observation in the note to the previous exercise.

Chapter 3
Completeness of the Classical Spaces

I often like women. I like their unconventionality. I like their completeness.

Virginia Woolf. A Room of One's Own

This chapter is mostly devoted to proving the completeness of the important spaces of multivariate calculus and basic functional analysis. Applications of completeness will be studied in chapter 6.

3.1 Coordinate Spaces and Normed Sequence Spaces

A basic theorem of fundamental analysis is Cauchy's principle of convergence, which asserts that a sequence of real numbers is convergent if and only if it satisfies Cauchy's condition. In other words $\mathbb{R}$ is a complete metric space when endowed with its usual metric $|s - t|$. This key fact of analysis is the starting point of all deliberations on the completeness of specific metric spaces.

3.1.1 Completeness of $\mathbb{R}^n$

We want to prove that $\mathbb{R}^n$ is complete. As it stands this is meaningless until we specify what metric we are considering, and we have already seen a large range of metrics, namely $d_p(x, y) = \|x - y\|_p$, for any p in the range $1 \le p \le \infty$. These are the only metrics we shall consider for $\mathbb{R}^n$. Fortunately it does not matter which one we use as we shall soon see.

We first show that convergence of a sequence in $\mathbb{R}^n$ is equivalent to convergence of the corresponding sequences of coordinates in $\mathbb{R}$, irrespective of which of these metrics is used. As always $\mathbb{R}$ has its usual metric $|s - t|$. We shall have to consider sequences of coordinate vectors. Since a coordinate vector is itself a sequence the

© The Author(s), under exclusive license to Springer Nature Switzerland AG 2022
R. Magnus, *Metric Spaces*, Springer Undergraduate Mathematics Series,
https://doi.org/10.1007/978-3-030-94946-4_3

notation can be challenging and we resort to some notational devices to achieve some clarity.

Proposition 3.1 *Let* $x^{(i)} = (x_1^{(i)}, \ldots, x_n^{(i)})$ *for* $i = 1, 2, 3, \ldots$ *define a sequence of vectors in* $\mathbb{R}^n$ *and let* $y = (y_1, \ldots, y_n)$. *Then* $\lim_{i \to \infty} x^{(i)} = y$, *with respect to the metric* d_p, *if and only if* $\lim_{i \to \infty} x_k^{(i)} = y_k$ *for each of* $k = 1, 2, \ldots, n$.

Proof It is obvious that $\|x^{(i)} - y\|_\infty \to 0$ is equivalent to $x_k^{(i)} - y_k \to 0$ for $k = 1, \ldots, n$, since there are only finitely many coordinates. From the inequalities (5.3 exercise 1):

$$|x_k| \le \|x\|_\infty \le \|x\|_p \le n^{\frac{1}{p}} \|x\|_\infty$$

valid for $k = 1, \ldots, n$, $x = (x_1, \ldots, x_n) \in \mathbb{R}^n$ and $1 \le p < \infty$, we then deduce that $\|x^{(i)} - y\|_p \to 0$ if and only if $x_k^{(i)} - y_k \to 0$ for $k = 1, \ldots, n$.   □

Proposition 3.2 $\mathbb{R}^n$ *is complete with respect to the metric* d_p *(where* p *is any constant in the interval* $1 \le p \le \infty$).

Proof Let $(x^{(i)})_{i=1}^\infty$ be a Cauchy sequence in $\mathbb{R}^n$. Now

$$|x_k^{(i)} - x_k^{(j)}| \le \|x^{(i)} - x^{(j)}\|_p,$$

so that for each of $k = 1, 2, .., n$ the real number sequence $(x_k^{(i)})_{i=1}^\infty$ is a Cauchy sequence. Hence for each of $k = 1, 2, .., n$ there exists y_k, such that $\lim_{i \to \infty} x_k^{(i)} = y_k$. But then we obtain $\lim_{i \to \infty} x^{(i)} = y$ where y is the vector $(y_1, y_2, \ldots, y_n)$.   □

3.1.2 Completeness of ℓ^p

The spaces ℓ^p are also complete, but the proofs are less simple owing to the infinite number of coordinates. If $(x^{(i)})_{i=1}^\infty$ is a convergent sequence in ℓ^p with limit y, then it is the case that $\lim_{i \to \infty} x_k^{(i)} = y_k$ for each of $k = 1, 2, 3, \ldots$. But the converse does not hold. Something further is required beyond mere convergence of each coordinate sequence.

Proposition 3.3 *The space* ℓ^p *is complete.*

Proof Let $(x^{(i)})_{i=1}^\infty$ be a Cauchy sequence in ℓ^p. The inequalities

$$|x_k| \le \|x\|_\infty \le \|x\|_p$$

hold for infinite sequences, as well as finite ones. We conclude that the real number sequences $(x_k^{(i)})_{i=1}^\infty$ $(k = 1, 2, \ldots)$ are Cauchy sequences, and each has therefore a finite limit y_k. Set $y = (y_1, y_2, \ldots)$.

The vector y is a candidate for the limit $\lim_{i \to \infty} x^{(i)}$. We need to show both that $y \in \ell^p$ and that $\|x^{(i)} - y\|_p \to 0$.

Recall that a Cauchy sequence is always bounded. There exists K, such that $\|x^{(i)}\|_p \leq K$ for all i. We consider separately the cases $p < \infty$ and $p = \infty$.

The case $p < \infty$. By Minkowski's inequality we have, for each n and i

$$\left(\sum_{k=1}^{n} |y_k|^p \right)^{\frac{1}{p}} \leq \left(\sum_{k=1}^{n} |x_k^{(i)} - y_k|^p \right)^{\frac{1}{p}} + \left(\sum_{k=1}^{n} |x_k^{(i)}|^p \right)^{\frac{1}{p}}$$

$$\leq \left(\sum_{k=1}^{n} |x_k^{(i)} - y_k|^p \right)^{\frac{1}{p}} + K$$

Let $i \to \infty$. The finite sum on the right tends to 0 and we find that

$$\left(\sum_{k=1}^{n} |y_k|^p \right)^{\frac{1}{p}} \leq K.$$

Since this holds for all n we conclude that

$$\left(\sum_{k=1}^{\infty} |y_k|^p \right)^{\frac{1}{p}} \leq K.$$

That is, $y \in \ell^p$.

Let $\varepsilon > 0$. There exists N, such that $\|x^{(i)} - x^{(j)}\|_p < \varepsilon$ for all $i \geq N$ and $j \geq N$. Fix natural numbers $i \geq N$ and n. Then for all $j \geq N$ we have

$$\left(\sum_{k=1}^{n} |x_k^{(i)} - x_k^{(j)}|^p \right)^{\frac{1}{p}} < \|x^{(i)} - x^{(j)}\|_p < \varepsilon.$$

Let $j \to \infty$ and deduce, for all $i \geq N$ and all n, that

$$\left(\sum_{k=1}^{n} |x_k^{(i)} - y_k|^p \right)^{\frac{1}{p}} \leq \varepsilon.$$

Now let $n \to \infty$ and deduce that for all $i \geq N$ we have $\|x^{(i)} - y\|_p \leq \varepsilon$. Hence $x^{(i)} \to y$ in ℓ^p.

The case $p = \infty$. There exists K, such that $\|x^{(i)}\|_\infty < K$ for all i. Hence, for each k we find

$$|y_k| \leq |y_k - x_k^{(i)}| + |x_k^{(i)}| < |y_k - x_k^{(i)}| + K.$$

Let $i \to \infty$ and deduce that $|y_k| \leq K$ for each k. Hence $y \in \ell^\infty$.

Let $\varepsilon > 0$. There exists N, such that $\|x^{(i)} - x^{(j)}\|_\infty < \varepsilon$ for all $i \geq N$ and $j \geq N$. For all k, and all $i \geq N$ and $j \geq N$, we have $|x_k^{(i)} - x_k^{(j)}| < \varepsilon$. We let $j \to \infty$ and deduce that $|x_k^{(i)} - y_k| \leq \varepsilon$ for all k and all $i \geq N$. Hence for all $i \geq N$ we have $\|x^{(i)} - y\|_\infty \leq \varepsilon$. This shows that $x^{(i)} \to y$ in ℓ^∞.   □

A complete normed space is called a Banach space. A very large part of functional analysis is taken up by the study of Banach spaces. We shall see many more examples in the ensuing pages.

3.1.3 Exercises

1. Show that all closed subsets of $\mathbb{R}^n$ are complete.

   Note. The intended meaning is that if $\mathbb{R}^n$ is equipped with the metric $d_p(x, y) = \|x - y\|_p$ (for some p in the range $1 \leq p \leq \infty$), and if A is a closed subset of $\mathbb{R}^n$, then (A, d_p) is a complete metric space. This provides a vast array of complete metric spaces that are useful in multivariate calculus.

2. The space ℓ^∞ has two important linear subspaces (one of which was introduced in section 2.4 exercise 13). In Banach's notation (convention requires here the lower case letters) they are

   i) c : the space of all convergent numerical sequences.
   ii) c_0 : the space of all numerical sequences with the limit 0.

   Show the following:

   a) c and c_0 are closed sets in ℓ^∞.
      Note. Often one says that they are closed *subspaces* of ℓ^∞, but 'subspace' here refers to linear subspace, not subspace in the sense of metric spaces.
   b) c and c_0 are Banach spaces.
   c) The linear functional $\lambda : c \to \mathbb{R}$, given by $\lambda(a) = \lim_{n\to\infty} a_n$, is continuous and has norm 1.

3.2 Product Spaces

Having considered the coordinate spaces $\mathbb{R}^n$ and their generalisations ℓ^p we turn to the more general case of products of metric spaces. Just as for $\mathbb{R}^n$ there is a choice of metrics.

3.2.1 Finitely Many Factors

A sequence $\left((x_n, y_n)\right)_{n=1}^{\infty}$ in the product of two spaces $X \times Y$ should converge to a point (x, y) if and only if $x_n \to x$ and $y_n \to y$. This is *coordinate-wise convergence*. Several commonly used distinct metrics on $X \times Y$ lead to coordinate-wise convergence, and therefore are equivalent metrics since the same sets are open irrespective of the metric used. For example, in addition to the default metric

$$D\left((x, y), (x', y')\right) = \max\left(d(x, x'), d(y, y')\right)$$

we can name the often used metrics

$$D_1\left((x, y), (x', y')\right) = d(x, x') + d(y, y')$$

$$D_2\left((x, y), (x', y')\right) = \sqrt{d(x, x')^2 + d(y, y')^2}$$

The objective is to define a metric in product spaces with an arbitrary number of factors so that convergence shall be coordinate-wise convergence. For the product X^n, it is usually simplest to use the metric

$$D\left((x_1, \ldots, x_n), (y_1, \ldots, y_n)\right) = \max_{1 \le k \le n}\left(d(x_k, y_k)\right).$$

The factors can be distinct spaces (X_k, d_k). For the product $\Pi_{k=1}^{n} X_k$ we can use the metric

$$D\left((x_1, \ldots, x_n), (y_1, \ldots, y_n)\right) = \max_{1 \le k \le n}\left(d_k(x_k, y_k)\right).$$

The reader should check that these metrics achieve the stated aim, that convergence is coordinate-wise.

3.2.2 Infinitely Many Factors

Let (X, d) be a metric space. Our aim is to define a metric on $X^{\mathbb{N}+}$ (the set of all sequences $(x_n)_{n=1}^{\infty}$ in X), so that convergence of a sequence in $X^{\mathbb{N}+}$ is coordinate-wise convergence. Only then do we call it a product space. This means that a sequence of sequences,

$$x^{(i)} = (x_1^{(i)}, x_2^{(i)}, \ldots), \quad (i = 1, 2, 3, \ldots),$$

converges to a sequence $y = (y_1, y_2, \ldots)$ if and only if

$$\lim_{i \to \infty} x_n^{(i)} = y_n \quad \text{for} \quad n = 1, 2, 3, \ldots.$$

This is not the case for the normed spaces ℓ^p, for which coordinate-wise convergence is weaker than convergence in the metric $\|x - y\|_p$.

One way to accomplish this aim is to use the metric (compare 1.4 exercise 3)

$$D(x, y) = \sum_{n=1}^{\infty} \frac{2^{-n} d(x_n, y_n)}{1 + d(x_n, y_n)}, \quad \left(x = (x_n)_{n=1}^{\infty}, \ y = (y_n)_{n=1}^{\infty}\right)$$

Proposition 3.4 *Convergence in the metric D is equivalent to coordinate-wise convergence.*

Proof We show first that the mapping

$$\pi_m : X^{\mathbb{N}+} \to X, \quad \pi_m\left((x_n)_{n=1}^{\infty}\right) = x_m$$

that assigns to each sequence $x = (x_n)_{n=1}^{\infty}$ its mth coordinate x_m, is continuous. Firstly it is clear that for all x and y in $X^{\mathbb{N}+}$, and all m, we have

$$\frac{2^{-m} d(x_m, y_m)}{1 + d(x_m, y_m)} < D(x, y). \tag{3.1}$$

Moreover the inverse of the function $x/(1 + x)$ is continuous at 0. Hence, for all $\varepsilon > 0$ and m there exists $\delta > 0$, such that, for all x and y in $X^{\mathbb{N}+}$, if $D(x, y) < \delta$ then $d(x_m, y_m) < \varepsilon$. This implies that the coordinate function π_m is continuous.

We can conclude the following about convergence in $X^{\mathbb{N}+}$. If the sequence $(x^{(i)})_{i=1}^{\infty}$ in $X^{\mathbb{N}+}$ is convergent and its limit is $a \in X^{\mathbb{N}+}$, then for all m, the sequence $(x_m^{(i)})_{i=1}^{\infty}$ in X, the sequence of mth coordinates, converges to a_m in X.

Next the converse. Suppose that the sequence $(x^{(i)})_{i=1}^{\infty}$ in $X^{\mathbb{N}+}$ has the property that for each m the sequence $(x_m^{(i)})_{i=1}^{\infty}$ is convergent in X. Set

$$a_m = \lim_{i \to \infty} x_m^{(i)}, \quad (m = 1, 2, 3, \ldots)$$

$$a = (a_m)_{m=1}^{\infty}.$$

We shall show that $\lim_{i \to \infty} x^{(i)} = a$.

Let $\varepsilon > 0$. Choose L so that

$$\sum_{k=L+1}^{\infty} 2^{-k} < \frac{\varepsilon}{2}.$$

Now with the chosen L there exists N, such that $i \geq N$ implies

$$\sum_{k=1}^{L} \frac{2^{-k} d(x_k^{(i)}, a_k)}{1 + d(x_k^{(i)}, a_k)} < \frac{\varepsilon}{2}.$$

Then $i \geq N$ implies $D(x^{(i)}, a) < \varepsilon$. $\qquad\square$

The metrics defined above for the product spaces X^n and $X^{\mathbb{N}_+}$ generalise in an obvious way to the case of distinct factors. Thus for a sequence (X_n, d_n), $(k = 1, 2, \ldots)$, of metric spaces we can define metrics

$$D(x, y) = \max_{1 \leq k \leq n} d_k(x_k, y_k), \quad \left(x, \ y \in \prod_{k=1}^{n} X_k\right)$$

or

$$D'(x, y) = \sum_{k=1}^{\infty} \frac{2^{-k} d_k(x_k, y_k)}{1 + d_k(x_k, y_k)}, \quad \left(x, \ y \in \prod_{k=1}^{\infty} X_k\right).$$

Proposition 3.5 *Let (X, d) be a complete metric space. Then (X^n, D) is complete for each $n \in \mathbb{N}_+$. Furthermore $(X^{\mathbb{N}_+}, D)$ is complete.*

Proof For X^n the proof is the same as for $\mathbb{R}^n$. Consider the case $X^{\mathbb{N}_+}$. Let $(x^{(i)})_{i=1}^{\infty}$ be a Cauchy sequence in $X^{\mathbb{N}_+}$ (using the metric D). Using the inequality (3.1) we see that for each m the sequence $(x_m^{(i)})_{i=1}^{\infty}$ is a Cauchy sequence in X, and therefore convergent in X. But then $(x^{(i)})_{i=1}^{\infty}$ is convergent in $X^{\mathbb{N}_+}$. $\qquad\square$

The same proof can be used for the case of distinct factors $\prod_{k=1}^{\infty} X_k$.

3.2.3 The Space $2^{\mathbb{N}_+}$ and the Cantor Set

An important example of a product space is the space

$$2^{\mathbb{N}_+} := \{0, 1\}^{\mathbb{N}_+},$$

comprising all sequences of the digits 0 and 1. The factor $X := \{0, 1\}$ is a discrete space; all four of its subsets are open sets. A suitable metric on X would have $d(0, 1) = 1$ for example.

A sequence in a discrete space is convergent if and only if it is eventually constant. Therefore a sequence $(x^{(i)})_{i=1}^{\infty}$, in $2^{\mathbb{N}_+}$ is convergent if and only if its coordinates "freeze" from left to right, with increasing i. More precisely, if and only if, for each p there exists m, such that for $i \geq m$ the coordinates $x_1^{(i)}, \ldots, x_p^{(i)}$ become fixed.

Cantor's middle thirds set was defined in section 1.3.

Proposition 3.6 *The product space $2^{\mathbb{N}+}$ is homeomorphic to Cantor's middle thirds set B.*

Proof Cantor's middle thirds set B is viewed as a subspace of $\mathbb{R}$, with the metric $|s - t|$. It comprises all points x in the interval $[0, 1]$ that may be written as a ternary fraction $x = (0.a_1a_2a_3 \ldots)_3$ in which each digit is 0 or 2. This representation is unique if it exists.

For $x = (0.a_1a_2a_3 \ldots)_3 \in B$ we define $f(x) \in 2^{\mathbb{N}+}$ by

$$f(x) = (c_1, c_2, c_3, \ldots)$$

where $c_k = 0$ if $a_k = 0$ and $c_k = 1$ if $a_k = 2$. It is obvious that $f : B \to 2^{\mathbb{N}+}$ is bijective. We shall show that f is a homeomorphism.

It is easy to see that f^{-1} is continuous. We use the characterisation of continuity using convergent sequences (section 2.3.4). Suppose that

$$(c_1^{(i)}, c_2^{(i)}, c_3^{(i)}, \ldots) \to (c_1, c_2, c_3, \ldots)$$

(as $i \to \infty$) in $2^{\mathbb{N}+}$, and let

$$a_n^{(i)} = 2c_n^{(i)}, \quad a_n = 2c_n.$$

Then we have real number convergence

$$(0.a_1^{(i)} a_2^{(i)} a_3^{(i)} \ldots.)_3 \to (0.a_1a_2a_3 \ldots)_3$$

in the usual metric of $\mathbb{R}$, because the digits are freezing from left to right.

It is slightly trickier to see that f is continuous. Let $x_i \in B$, $x \in B$ and assume that $\lim_{i \to \infty} x_i \to x$ in the metric $|s - t|$. We want to show that $\lim_{i \to \infty} f(x_i) = f(x)$, where the limit is understood in the metric of $2^{\mathbb{N}+}$.

Let

$$x_i = (0.a_1^{(i)} a_2^{(i)} a_3^{(i)} \ldots.)_3, \quad x = (0.a_1a_2a_3 \ldots.)_3,$$

in the ternary system, where each digit is 0 or 2. Fix an index i and suppose that x_i and x differ first at the nth digit. Then $|a_n^{(i)} - a_n| = 2$ and so

$$|x_i - x| = \left| \sum_{j=n}^{\infty} \frac{a_j^{(i)} - a_j}{3^j} \right| \geq \frac{|a_n^{(i)} - a_n|}{3^n} - \left| \sum_{j=n+1}^{\infty} \frac{a_j^{(i)} - a_j}{3^j} \right|$$

$$\geq \frac{2}{3^n} - \sum_{j=n+1}^{\infty} \frac{2}{3^j} = \frac{1}{3^n}.$$

Of course n depends on i; let us write $n = n(i)$. We have just shown that

$$\frac{1}{3^{n(i)}} \leq |x_i - x|.$$

Since $x_i \to x$ in $\mathbb{R}$ we see that $n(i) \to \infty$ as $i \to \infty$. The digits $a_j^{(i)}$, ($j = 1, 2, 3, \ldots$), freeze from left to right with increasing i. In other words $f(x_i)$ converges to $f(x)$ in the space $2^{\mathbb{N}_+}$. □

3.2.4 Subspaces of Complete Spaces

The conclusion of the previous paragraph, that Cantor's middle thirds set B is homeomorphic to the complete metric space $2^{\mathbb{N}_+}$, is not an argument for the completeness of B. It is worth repeating that completeness is not a topological property. That B is complete follows instead almost trivially from the fact that it is closed in $\mathbb{R}$.

More generally, the available supply of complete metric spaces is vastly increased by the following result, already hinted at in several places, but never formally stated.

Proposition 3.7 Let (X, d) be a metric space and let $A \subset X$. Then:

1) If (A, d) is complete then A is a closed set in X.
2) If (X, d) is complete and A a closed set then (A, d) is complete.

Proof
1) Suppose that (A, d) is complete. Let $(x_n)_{n=1}^{\infty}$ be a sequence in A that converges in (X, d) and let y be its limit. Now $(x_n)_{n=1}^{\infty}$ is a Cauchy sequence with respect to the metric d, and, as it lies in A it is a Cauchy sequence in (A, d). Since (A, d) is complete, the sequence $(x_n)_{n=1}^{\infty}$ converges in (A, d). But the metrics of (A, d) and (X, d) are the same, and limits are unique, so that $y \in A$. In other words A contains the limits of all its convergent sequences, and therefore A is closed in X.

2) Suppose that X is complete and A closed. Let $(x_n)_{n=1}^\infty$ be a Cauchy sequence in (A, d). Then obviously $(x_n)_{n=1}^\infty$ is also a Cauchy sequence in (X, d), hence converges in X. Let its limit be y. Then $y \in A$ because A is closed in X. Hence $(x_n)_{n=1}^\infty$ converges also in A.   □

3.2.5   Exercises

1. Let $1 \le p \le \infty$. Show that the mapping $J : \ell^p \to \mathbb{R}^{\mathbb{N}+}$, the embedding of ℓ^p into $\mathbb{R}^{\mathbb{N}+}$, defined by $J(x) = x$, is continuous but its inverse (defined on its range) is not continuous.

   Note. Injective mappings, especially those that fix the points of their domains, are sometimes called embeddings, but the use of the term can be confusing. Strictly speaking, the mapping J is not a *topological embedding* as its inverse is not continuous. It might be reasonable to call it an embedding in a purely set theoretical sense; or in the purely algebraic sense of real vector spaces. Whether an injective mapping is an embedding depends on the *category*; topological spaces, sets and real vector spaces being three different categories (in the sense of *category theory* which deals with questions of this kind). Despite this the term 'embedding' is traditionally attached to certain continuous injective linear mappings whose inverses are not continuous, such as the *Sobolev embeddings*, central to the theory of partial differential equations.

2. The space $2^{\mathbb{N}+}$ (and therefore also the Cantor middle thirds set) is homeomorphic to an ultrametric space. For two sequences a and b comprising the digits 0 and 1 we set $d(a, b) = 2^{-n}$, where n is the lowest place number at which a and b differ.

   a) Show that d is an ultrametric.
   b) Show that d is equivalent to any of the usual product metrics.
      Hint. Forget the product metrics. Just consider what convergence of a sequence, $a^{(i)} \to b$, means in the new metric.

3. Show that $2^{\mathbb{N}+} \times 2^{\mathbb{N}+}$ is homeomorphic to $2^{\mathbb{N}+}$.
4. Go even further! Show that $(2^{\mathbb{N}+})^{\mathbb{N}+}$ is homeomorphic to $2^{\mathbb{N}+}$.
5. A very general way to impose a metric on the product $X^{\mathbb{N}+}$ is to write

$$D'(x, y) = \left\| \left(\beta_n f(d(x_n, y_n)) \right)_{n=1}^\infty \right\|_p$$

   where $1 \le p \le \infty$, $(\beta_n)_{n=1}^\infty \in \ell^p$, and f is a function satisfying the conditions of 1.1 exercise 5, that is additionally continuous at 0 and bounded. In the case $p = \infty$ we require that $\beta_n \to 0$.

   The metric $D(x, y)$ that we used in the text corresponded to $p = 1$, $f(t) = t/(t + 1)$, $\beta_n = 2^{-n}$. Show that D' is equivalent to D.

   Hint. Show that the criterion for convergence of a sequence is the same for both.

Note. Which metric one uses is a matter of convenience. Arguably it may be best to use $p = \infty$, $\beta_n = 1/n$ and $f(t) = \min(t, 1)$. This gives

$$D'(x, y) = \max_{1 \leq n < \infty} \frac{1}{n} \min(d(x_n, y_n), 1).$$

Note that we metrise the product of different spaces $\prod_{n=1}^{\infty} X_n$ by replacing $d(x_n, y_n)$ by $d_n(x_n, y_n)$ in these formulas.

6. In this exercise we need the concept of a topological space and that of continuity for mappings whose domain and codomain are topological spaces. We also need the notion of subbase for a topology. See section 2.6 and its exercises.

   The plethora of different metrics on a product space, all of which give rise to the same open sets, suggests that it should be possible, and even desirable, to define the topology of a product space without using a metric. This is in fact the preferable way to treat product spaces because it applies equally to uncountable products.

   Let $(X_i)_{i \in I}$ be a family of topological spaces. For each i let π_i be the coordinate function

   $$\pi_i : \prod_{k \in I} X_k \to X_i, \qquad \pi_i\big((x_k)_{k \in I}\big) = x_i.$$

   The product topology is defined to be the minimum topology (or the coarsest topology, as it is usually called by topologists) on $\prod_{k \in I} X_k$, that makes all the coordinate functions continuous.

   a) Show that the sets $\pi_i^{-1}(U)$, as i ranges over I and, for each i, U ranges over all open sets in X_i, constitute a subbase for the product topology.
   b) Show that a non-empty set $\mathbf{F} \subset \prod_{i \in I} X_i$ is open if and only if it satisfies the following condition: for each $\mathbf{a} \in F$ there exists a finite set $\{i_1, \ldots, i_n\} \subset I$ and sets $U_k \subset X_{i_k}$, for $k = 1, \ldots, n$, open in X_{i_k}, such that

   $$\mathbf{a} \in \bigcap_{k=1}^{n} \pi_{i_k}^{-1}(U_k) \subset \mathbf{F}.$$

   Note. It is helpful to have a mental picture of the set $\pi_i^{-1}(U)$. It is the set of all families $(x_k)_{k \in I}$ for which x_i is restricted to U but all other coordinates are unrestricted. Such a set is called a cylinder set in the product space. Finite intersections of such sets are also called cylinder sets. The geometrical terminology is perhaps not so helpful; the unrestricted coordinates in the families making up a cylinder set are perhaps likened to the generating lines of a cylinder.

   c) Let Y be a topological space and let $f : Y \to \prod_{i \in I} X_i$. Show that f is continuous if and only if all the mappings $\pi_i \circ f : Y \to X_i$, $(i \in I)$, (the coordinate functions of f) are continuous.

d) Show that if all the spaces X_i are Hausdorff, then the product space $\prod_{i \in I} X_i$ is also Hausdorff.

e) Show that in the countable case, $\prod_{k=1}^{\infty} X_k$, where each X_k is a metric space, the product topology as just defined has the same open sets as result from any of the product metrics discussed in exercise 5.

7. Let X be a metric space and let $(U_n)_{n=1}^{\infty}$ be a sequence of non-empty subsets of X. Viewing each U_n as a space in its own right we form the product

$$W = \prod_{n=1}^{\infty} U_n.$$

Let Δ be the subset of W consisting of all constant sequences. There is an obvious bijection

$$G : \bigcap_{n=1}^{\infty} U_n \to \Delta$$

that maps $y \in \bigcap_{n=1}^{\infty} U_n$ to the constant sequence $(x_n)_{n=1}^{\infty}$ with $x_n = y$.

a) Show that Δ is a closed set in W.
b) Show that G is a homeomorphism.

8. Let X be a complete metric space.

a) Let A be a non-empty open set in X. Show that A is homeomorphic to a complete metric space.
   Hint. One way is to define

$$g : A \to X \times \mathbb{R}, \quad g(x) = (x, 1/d(x, A^c)).$$

b) Let $(A_n)_{n=1}^{\infty}$ be a sequence of open sets in X. Show that their intersection, $\bigcap_{n=1}^{\infty} A_n$, if non-empty, is homeomorphic to a complete metric space.
   Hint. Refer to the previous exercise.

9. Show that the space of all irrational numbers is homeomorphic to a complete metric space.
   Hint. Use the preceding exercise. The result may seem surprising. Nevertheless, a practical homeomorphism, mapping the space of irrationals to the product space $\mathbb{N}_+^{\mathbb{N}}$, can be realised using continued fractions.

3.3 Spaces of Continuous Functions

If (M, d) is a metric space we can study the space $C(M)$ consisting of all continuous functions $f : M \to \mathbb{R}$. This is also a vector space. If f and g are in $C(M)$ and $\lambda \in \mathbb{R}$ then the sum $f + g$ and product λf are continuous functions.

It is easiest to show this using sequences. Let $(x_n)_{n=1}^{\infty}$ be a convergent sequence in M, with limit a. Then $f(x_n) \to f(a)$ and $g(x_n) \to g(a)$; so we conclude that $f(x_n) + g(x_n) \to f(a) + g(a)$ and $\lambda f(x_n) \to \lambda f(a)$.

We would like to make $C(M)$ into a normed space by using the norm

$$\|f\|_{\infty} = \sup_{x \in M} |f(x)|.$$

The problem is that continuous functions on M are not automatically bounded; so $\|f\|_{\infty}$ may be ∞. We know by fundamental analysis that continuous functions on the space $[a, b]$ are bounded. The question then arises: for what spaces M are continuous functions $f : M \to \mathbb{R}$ necessarily bounded? This question will be answered later, but meanwhile we can restrict our attention to those functions on M that are both continuous and bounded. They constitute a normed vector space $C_B(M)$ with the norm $\|f\|_{\infty}$.

3.3.1 Uniform Convergence

Let M be a set and let $f_n : M \to \mathbb{R}$ for each $n \in \mathbb{N}_+$. We do not yet assume a metric in M so continuity is not yet involved. The assertion $\lim_{n \to \infty} f_n = g$ may be interpreted in two different ways:

a) Pointwise convergence. For each $x \in M$ the real number sequence $(f_n(x))_{n=1}^{\infty}$ converges to $g(x)$.

b) Uniform convergence. For each $\varepsilon > 0$ there exists N, such that for all $n \geq N$ and all $x \in M$ we have $|f_n(x) - g(x)| < \varepsilon$.

It is informative to write the definitions of pointwise convergence and uniform convergence as sentences of first order logic. They involve the same assertion, preceded by the same four quantifiers—only their order is different. The first is pointwise, the second uniform convergence:

$$(\forall \varepsilon > 0)(\forall x \in M)(\exists N \in \mathbb{N})(\forall n \in \mathbb{N})(n \geq N \Rightarrow |f_n(x) - g(x)| < \varepsilon)$$

$$(\forall \varepsilon > 0)(\exists N \in \mathbb{N})(\forall n \in \mathbb{N})(\forall x \in M)(n \geq N \Rightarrow |f_n(x) - g(x)| < \varepsilon)$$

Clearly uniform convergence implies pointwise convergence. However the converse is untrue.

The importance of uniform convergence is mainly due to the following proposition.

Proposition 3.8 *Let (M, d) be a metric space, and let $(f_n)_{n=1}^{\infty}$ be a sequence of continuous functions from M to $\mathbb{R}$. Assume that f_n converges uniformly to a function $g : M \to \mathbb{R}$. Then g is continuous.*

Proof Let $a \in M$. We shall show that g is continuous at a.

Let $\varepsilon > 0$. There exists N, such that $|f_n(x) - g(x)| < \varepsilon/3$ for all $n \geq N$ and all $x \in M$. Since f_N is continuous at a, there exists $\delta > 0$, such that $|f_N(x) - f_N(a)| < \varepsilon/3$ if $d(x, a) < \delta$. If now $d(x, a) < \delta$ we obtain:

$$|g(x) - f_N(x)| < \varepsilon/3, \quad |f_N(x) - f_N(a)| < \varepsilon/3, \quad |f_N(a) - g(a)| < \varepsilon/3$$

and therefore $|g(x) - g(a)| < \varepsilon$.   □

We note, for possible future reference, that the argument, showing that g is continuous at a, does not require continuity of the functions f_n except at the point a.

We consider next the normed space $C_B(M)$ with $\|f\|_{\infty} = \sup_M |f(x)|$. We shall show that convergence in the metric determined by this norm is the same as uniform convergence of functions. This was previously observed in section 1.4 for spaces of bounded functions.

Suppose first that $\lim_{n \to \infty} f_n = g$ in the norm metric of $C_B(M)$. Let $\varepsilon > 0$. There exists N, such that

$$\sup_{x \in M} |f_n(x) - g(x)| < \varepsilon$$

for all $n \geq N$. But then $|f_n(x) - g(x)| < \varepsilon$ for all $x \in M$ and $n \geq N$. In other words f_n converges uniformly to g.

Conversely, suppose that $(f_n)_{n=1}^{\infty}$ is a sequence of functions in $C_B(M)$ such that $f_n \to g$ uniformly. We know, by proposition 3.8, that g is continuous and it is also bounded (for example, it differs from one of the functions f_n by less than 1). Hence $g \in C_B(M)$. This supplies a candidate function for the limit. We still need to prove convergence in norm.

Let $\varepsilon > 0$. There exists N, such that $|f_n(x) - g(x)| < \varepsilon$ for all $n \geq N$ and all $x \in M$. It is a key requirement that N can be chosen independently of x, a requirement fulfilled by uniform convergence. So we can conclude that

$$\sup_{x \in M} |f_n(x) - g(x)| \leq \varepsilon$$

for all $n \geq N$, that is,

$$\|f_n - g\|_{\infty} \leq \varepsilon.$$

This implies convergence in the norm.

In contrast to uniform convergence, it is not usually possible to capture pointwise convergence of functions with a metric on a space of functions.

Proposition 3.9 *The space $C_B(M)$ is complete.*

Proof Let $(f_n)_{n=1}^{\infty}$ be a Cauchy sequence in $C_B(M)$. Since for each $x \in M$ we have

$$|f_n(x) - f_m(x)| \leq \|f_n - f_m\|_{\infty},$$

we see that the numerical sequence $(f_n(x))_{n=1}^{\infty}$ is a Cauchy sequence. The limit $\lim_{n \to \infty} f_n(x)$ therefore exists for each $x \in M$; denote it by $g(x)$. The function g is a candidate for the limit $\lim_{n \to \infty} f_n$ in $C_B(M)$.

We only have to show that the limit $\lim_{n \to \infty} f_n(x) = g(x)$ is attained uniformly with respect to $x \in M$. This will also imply that g is continuous.

Let $\varepsilon > 0$. There exists N, such that $\|f_n - f_m\|_{\infty} < \varepsilon$ for all $m \geq N$ and all $n \geq N$. It follows that for all $x \in M$, all $n \geq N$ and all $m \geq N$, we have $|f_n(x) - f_m(x)| < \varepsilon$. Fix x and let $m \to \infty$. We find that $|f_n(x) - g(x)| \leq \varepsilon$. This holds for all $x \in M$ and $n \geq N$. This says that $f_n \to g$ uniformly. □

Exercise We omitted to show in the proof that the function g was bounded. Rectify this. Show that in general a uniform limit of bounded functions on a set M is a bounded function.

3.3.2 Series in Normed Spaces

Let X be a normed vector space. We recall that the metric is given by $d(u, v) = \|u - v\|$. We recall also that if X is complete with this metric then we call X a Banach space. We are concerned here with the interplay of algebraic operations, in particular, addition of vectors, with limiting operations.

If we have a sequence of vectors $(u_n)_{n=1}^{\infty}$ in X we can form the sequence of finite sums, or partial sums as they are called:

$$w_n := \sum_{k=1}^{n} u_k, \quad (n = 1, 2, 3, \ldots)$$

but now we wish to study, and make sense of, the *infinite series*

$$\sum_{k=1}^{\infty} u_k$$

using the same procedure as is used to define the sum of an infinite series of real numbers. We say that the series $\sum_{n=1}^{\infty} u_n$ is convergent, and that its sum is the vector $v \in X$, if

$$\lim_{n \to \infty} w_n = v,$$

where $w_n = \sum_{k=1}^{n} u_k$, $n = 1, 2, 3, \ldots$.

The most important results require completeness of X.

Proposition 3.10 *Let X be a Banach space. Let $(u_n)_{n=1}^{\infty}$ be a sequence of vectors and assume that the numerical series of positive terms, $\sum_{n=1}^{\infty} \|u_n\|$, is convergent. Then the series $\sum_{n=1}^{\infty} u_n$ is convergent and we have an infinite version of the triangle inequality:*

$$\left\| \sum_{n=1}^{\infty} u_n \right\| \leq \sum_{n=1}^{\infty} \|u_n\|$$

Proof Let $\varepsilon > 0$. Since $\sum_{n=1}^{\infty} \|u_n\|$ is convergent there exists N, such that

$$\sum_{k=m+1}^{n} \|u_k\| < \varepsilon$$

for all m and n that satisfy $N \leq m \leq n$. If $w_n = \sum_{k=1}^{n} u_k$ and $N \leq m \leq n$ we have

$$\|w_n - w_m\| = \left\| \sum_{k=m+1}^{n} u_k \right\| \leq \sum_{k=m+1}^{n} \|u_k\| < \varepsilon$$

that is, $(w_n)_{n=1}^{\infty}$ is a Cauchy sequence in X. But then the limit $v = \lim_{n \to \infty} w_n$ exists. The triangle inequality gives

$$\left\| \sum_{k=1}^{n} u_k \right\| \leq \sum_{k=1}^{n} \|u_k\| \leq \sum_{k=1}^{\infty} \|u_k\|.$$

Now the norm is a continuous function on X. We may let $n \to \infty$ and deduce

$$\left\| \sum_{k=1}^{\infty} u_k \right\| \leq \sum_{k=1}^{\infty} \|u_k\|.$$

$\square$

When the condition that $\sum_{n=1}^{\infty} \|u_n\|$ is convergent is satisfied we say that the series $\sum_{n=1}^{\infty} u_n$ is *absolutely convergent*, in direct analogy to the case of numerical series.

3.3.3 The Weierstrass M-Test

Proposition 3.10 is applicable to series in $\mathbb{R}^n$ (also $\mathbb{C}$, which as a metric space is the same as $\mathbb{R}^2$). It is also useful in infinite-dimensional Banach spaces, such as ℓ^p and $C_B(M)$.

A series in $C_B(M)$ is of the form $\sum_{n=1}^{\infty} f_n$ where $f_n : M \to \mathbb{R}$ is bounded and continuous for each n. The proposition tells us that if the numerical series $\sum_{n=1}^{\infty} \| f_n \|_\infty$ is convergent then the function series $\sum_{n=1}^{\infty} f_n$ converges uniformly with respect to the set M. This is essentially the Weierstrass M-test familiar from fundamental analysis.

The usual Weierstrass M-test is as follows. Let M_n be numbers (in spite of the notational conflict with 'the set M' we use the traditional notation that probably explains the name), such that $|f_n(x)| \le M_n$ for each $x \in M$ and each n. We assume that $\sum_{n=1}^{\infty} M_n$ is convergent. Then the series $\sum_{n=1}^{\infty} f_n$ converges uniformly with respect to $x \in M$.

Obviously if $|f_n(x)| \le M_n$ for each $x \in M$ and each n, then $\| f_n \|_\infty \le M_n$, and if $\sum_{n=1}^{\infty} M_n$ is convergent then $\sum_{n=1}^{\infty} \| f_n \|_\infty$ is convergent.

It is apparent that there is no restriction to continuous functions here. The test is therefore applicable to determine the absolute convergence of series in the space $B(M)$ of bounded functions on M.

3.3.4 The Spaces $C(\mathbb{R})$ and $C^p(\mathbb{R})$

If we do not restrict to bounded functions then it is not clear how to equip $C(\mathbb{R})$ with a norm. There is another approach. We can devise a metric which entails that a sequence $(f_n)_{n=1}^{\infty}$ converges to g when $f_n \to g$ uniformly on the interval $[-K, K]$ for every $K > 0$. The result is akin to a product space.

In fact there is an injective mapping

$$J : C(\mathbb{R}) \to \prod_{N=1}^{\infty} C[-N, N]$$

given by

$$J(f) = (f|_{[-N,N]})_{N=1}^{\infty}.$$

So we can view $C(\mathbb{R})$ as a subspace of $\prod_{N=1}^{\infty} C[-N, N]$. We can, for example, use the metric

$$d(f, g) = \sum_{N=1}^{\infty} 2^{-N} \min \left(\sup_{[-N,N]} |f - g|, 1 \right),$$

which makes J a topological embedding. We may replace the function $\min(t, 1)$ by other functions $\phi(t)$ with the right properties (see for example 3.2 exercise 5). A popular alternative is $t/(t + 1)$. Again 2^{-N} can be replaced by a positive number a_N provided $\sum_{N=1}^{\infty} a_N < \infty$. Then the metric is

$$d'(f, g) = \sum_{N=1}^{\infty} a_N \phi \left(\sup_{[-N,N]} |f - g| \right).$$

The important thing is that $\lim_{n \to \infty} f_n = g$ if and only if $f_n \to g$ uniformly on each set $[-N, N]$. With such a metric the space $C(\mathbb{R})$ is complete. In fact it may be regarded as a closed subset of $\prod_{N=1}^{\infty} C[-N, N]$.

We can include derivatives in the picture. We let $C^p(\mathbb{R})$ denote the space of all functions $f : \mathbb{R} \to \mathbb{R}$ that have continuous derivatives up to order p. We want to assign to the statement that $\lim_{n \to \infty} f_n = g$ the following meaning:

For each k, such that $0 \leq k \leq p$, and for each N, we have $f_n^{(k)} \to g^{(k)}$ uniformly in $[-N, N]$.

There is a big choice of metrics as usual. For example

$$d(f, g) = \sum_{k=0}^{p} \sum_{N=1}^{\infty} 2^{-N} \min \left(\sup_{[-N,N]} |f^{(k)} - g^{(k)}|, 1 \right).$$

Once again this space is complete. A small input from fundamental analysis, concerning uniform convergence, is needed.

Finally we allow infinitely many derivatives and make $C^{\infty}(\mathbb{R})$ into a complete metric space, in such a way that a sequence of C^{∞} functions $(f_k)_{k=1}^{\infty}$ converges to a C^{∞} function g if and only if the following is satisfied:

For all positive integers m and N we have $f_k^{(m)} \to g^{(m)}$ uniformly in the interval $[-N, N]$.

This is accomplished by the metric (to name just one example):

$$d(f, g) = \sum_{k=0}^{\infty} \sum_{N=1}^{\infty} 2^{-N-k} \min \left(\sup_{[-N,N]} |f^{(k)} - g^{(k)}|, 1 \right).$$

A variant of this is the space $C^{\infty}(A)$ where A is an open interval. Now we can use an increasing sequence $(K_n)_{n=1}^{\infty}$ of bounded closed subintervals A, such that $\bigcup_{n=1}^{\infty} K_n = A$, and write

$$d(f, g) = \sum_{k=0}^{\infty} \sum_{N=1}^{\infty} 2^{-N-k} \min \left(\sup_{K_N} |f^{(k)} - g^{(k)}|, 1 \right).$$

For example if $A =]a, b[$ (with finite a and b) we can use $K_n = [a + \frac{1}{n}, b - \frac{1}{n}]$.

The point of the bounded closed intervals K_n is simply this: that a continuous function is necessarily bounded on such an interval. How to proceed when the domain is not an interval of real numbers will be explained in a later chapter, when the concept of compactness has been introduced.

3.3.5 Exercises

1. Let X be the space of all continuous functions $f : [0, 1] \to \mathbb{R}$ with the metric

$$d(f, g) = \int_0^1 |f(t) - g(t)| \, dt.$$

   In other words we view X as a normed vector space with the norm $\|f\|_1 = \int |f|$. Show that X is not complete.

   Hint. Let $f_n(x) = 1$ if $0 \le x < \frac{1}{2}$, $f_n(x) = 0$ if $\frac{1}{2} + \frac{1}{n} \le x \le 1$ and let $f_n(x)$ be of the form $ax + b$ if $\frac{1}{2} \le x < \frac{1}{2} + \frac{1}{n}$. Here a and b are chosen to make f_n continuous. Show that the sequence $(f_n)_{n=1}^\infty$ is a Cauchy sequence in X that is not convergent.

2. Show that the vector space $B(M)$ of all bounded real functions on a set M is complete with the norm $\|f\|_\infty := \sup_M |f|$. A particular case is the sequence space ℓ^∞, which is $B(\mathbb{N}_+)$.

3. It was stated in the text that pointwise convergence of functions cannot usually be captured by a metric. Here is a case when it can. Let M be countable. Define a metric d in the space of all functions $f : M \to \mathbb{R}$, such that $f_n \to f$ pointwise if and only if $d(f_n, f) \to 0$.

4. The definitions of ' f_n converges pointwise to g' and ' f_n converges uniformly to g', formulated in sentences of first order logic, require four quantifiers. Using some set theory, the number of quantifiers can be drastically reduced, to one for pointwise convergence and to two for uniform convergence.

   Let M be a metric space and let $(f_k)_{k=1}^\infty$ be a sequence of real valued functions with domain M. Let g be a function on M. For each $\varepsilon > 0$ and each natural number n, we let

$$A_n^\varepsilon = \left\{ x \in M : |f_k(x) - g(x)| < \varepsilon \text{ for all } k \ge n \right\}.$$

   Now show the following:

   a) For each $\varepsilon > 0$, the sequence $(A_n^\varepsilon)_{n=1}^\infty$ of sets is increasing with respect to set inclusion.

   b) f_n converges to g pointwise if and only if, for each ε we have $\bigcup_{n=1}^\infty A_n^\varepsilon = M$.

b) f_n converges to g uniformly if and only if, for each ε there exists n, such that $A_n^\varepsilon = M$.

5. Show that a necessary condition for convergence of the series $\sum_{n=1}^\infty u_n$ of vectors in a normed space X, is that $\lim_{n\to\infty} u_n = 0$.

6. Let $e^{(i)}$ be the sequence $(t_n)_{n=1}^\infty$, where $t_n = 0$ for all $n \neq i$ but $t_i = 1$. Obviously $e^{(i)} \in \ell^p$ for all p, $(1 \leq p \leq \infty)$.

   a) Show that if $1 \leq p < \infty$ and $a = (a_n)_{n=1}^\infty \in \ell^p$ then

   $$a = \sum_{n=1}^\infty a_n e^{(n)}$$

   where convergence is understood to be in the metric of ℓ^p. Show that the series is not necessarily absolutely convergent.

   b) Exhibit an example showing that the formula of the previous item is not necessarily valid for $p = \infty$.

   c) Show that the formula $a = \sum_{n=1}^\infty a_n e^{(n)}$ is valid for a in the space c_0 (see 3.1 exercise 2 for the definition of this space).

7. In this series of exercises we shall identify all continuous linear functionals on the sequence space ℓ^p. Recall that the space of continuous linear functionals on a normed space X, when it carries the operator norm, is called the normed dual, or simply dual, of X. It is denoted by X^*.

   Let $1 < p < \infty$ and let q be the *conjugate* index to p, that is, the one that satisfies $(1/p) + (1/q) = 1$. In 2.4 exercise 4 we saw that every element $(\lambda_n)_{n=1}^\infty \in \ell^q$ determines a continuous linear functional Λ on ℓ^p by the prescription

   $$\Lambda(a) = \sum_{n=1}^\infty \lambda_n a_n, \quad \left(a = (a_n)_{n=1}^\infty \in \ell^p\right) \qquad (3.2)$$

   The following items are devoted to the converse. Let Λ be an element of the dual of ℓ^p. The object is to show that there exists $(\lambda_n)_{n=1}^\infty \in \ell^q$, such that (3.2) holds.

   a) Let $\lambda_n = \Lambda(e^{(n)})$ where $e^{(n)}$ is the sequence with 1 in the nth place and 0 in all other places. Show that the series $\sum_{n=1}^\infty \lambda_n a_n$ converges to $\Lambda(a)$ for all $a = (a_n)_{n=1}^\infty \in \ell^p$.

   Hint. See the previous exercise.

   b) Deduce that

   $$\left|\sum_{n=1}^\infty \lambda_n a_n\right| \leq \|\Lambda\| \left(\sum_{n=1}^\infty |a_n|^p\right)^{1/p}, \quad (a = (a_n)_{n=1}^\infty \in \ell^p) \qquad (3.3)$$

c) Show that $(\lambda_n)_{n=1}^{\infty} \in \ell^q$.

   Hint. Set $a_n = |\lambda_n|^{q-1} \operatorname{sgn} \lambda_n$ for $1 \leq n \leq N$, and $a_n = 0$ for $n \geq N+1$ in the inequality (3.3).

d) Prove that

$$\|\Lambda\| = \left(\sum_{k=1}^{n} |\lambda_k|^q \right)^{1/q}$$

The preceding items show that we can identify $(\ell^p)^*$ with ℓ^q, and the identification extends to the norm. The result is reciprocal; we similarly identify $(\ell^q)^*$ with ℓ^p. The case $p = 1$ is quite different.

e) Show that if $\Lambda \in (\ell^1)^*$ then there exists a *bounded sequence* $(\lambda_n)_{n=1}^{\infty}$ such that (3.2) holds (with $p = 1$). Show that we obtain an identification of $(\ell^1)^*$ with ℓ^∞.

f) Show that if $(\lambda_n)_{n=1}^{\infty} \in \ell^1$ then it defines a continuous linear functional on ℓ^∞ by the formula (3.2) (with $p = \infty$).

g) Recall the space c_0 of sequences with limit 0 (section 6.1). This is a closed subspace of ℓ^∞. Show that a linear functional on ℓ^∞ that is of the form (3.2), and which additionally is 0 on c_0, would necessarily be 0 on all of ℓ^∞.

Nevertheless, we cannot identify $(\ell^\infty)^*$ with ℓ^1. There exist continuous linear functionals on ℓ^∞ that cannot thus be represented. There exists a non-zero continuous linear functional on ℓ^∞ that is 0 on the whole of the subspace c_0, and, by the last item, this cannot have the said representation. This is a consequence of the Hahn-Banach theorem, which is not included in this text. Its proof requires the axiom of choice.

8. Show that the space $C_0(\mathbb{R})$ (of continuous functions that vanish at infinity, see 2.2 exercise 19), is complete in the supremum norm.

9. Let $f : \mathbb{R} \to \mathbb{R}$ be continuous and satisfy $\lim_{x \to \pm\infty} f(x) = 0$. Let $g_n(x) = f(x - n)$, $(n \in \mathbb{Z})$. Show that $\lim_{n \to \pm\infty} g_n = 0$ in the metric space $C(\mathbb{R})$ (which is not a normed space), but that this does not hold in the Banach space $C_0(\mathbb{R})$.

10. Show that the spaces $C^p(\mathbb{R})$ and $C^\infty(\mathbb{R})$ are complete in the metrics defined in the text.

    Hint. Recall some facts from fundamental analysis about what can be deduced when $f_n \to g$ uniformly and $f_n' \to h$ uniformly.

11. Let $g_n(x) = e^{-(x-n)^2}$, $(n \in \mathbb{N}_+)$. Show that $\lim_{n \to \infty} g_n = 0$ in the metric space $C^\infty(\mathbb{R})$.

12. The treatment of uniformly convergent sequences of functions in the text was entirely for real valued functions. We can consider functions $f : M \to Y$ where Y is a metric space. The general definition should be clear. The sequence of functions $f_n : M \to Y$, $(n = 1, 2, 3, \ldots)$, converges uniformly to a function $g : M \to Y$ when it satisfies the following condition:

*For all $\varepsilon > 0$ there exists N, such that $d(f_n(x), g(x)) < \varepsilon$ for all $n \geq N$
and all $x \in M$.*

Prove the analogue of proposition 3.8. Suppose that both M and Y are metric spaces, and suppose that $(f_n)_{n=1}^{\infty}$ is a sequence of continuous functions from M to Y, such that $f_n \to g$ uniformly. Then g is continuous.

3.4 ($\Diamond$) Rearrangements

Recall from analysis that a numerical series $\sum_{k=1}^{\infty} a_k$ is called *absolutely convergent* when the positive series $\sum_{k=1}^{\infty} |a_k|$ is convergent. On the other hand, the series $\sum_{k=1}^{\infty} a_k$ is called *conditionally convergent* when it is convergent but the series $\sum_{k=1}^{\infty} |a_k|$ is divergent.

A *permutation* of $\mathbb{N}_+$ is by definition a bijective mapping $\phi : \mathbb{N}_+ \to \mathbb{N}_+$. It acts on the series $\sum_{k=1}^{\infty} a_k$ by transforming it into the series $\sum_{k=1}^{\infty} a_{\phi(k)}$. This transformation is called a *rearrangement* of the initial series.

Riemann's theorem on rearrangements says that if a series $\sum_{k=1}^{\infty} a_k$ is conditionally convergent, then for any $t \in \mathbb{R} \cup \{-\infty, +\infty\}$, there is a rearrangement that has sum t. On the other hand we know that if a series is absolutely convergent then all its rearrangements have the same sum. From this we can also conclude that if all rearrangements of the series $\sum_{k=1}^{\infty} a_k$ are convergent then the series is absolutely convergent and all rearrangements have the same sum.

This two-fold classification of convergent series does not work for vector series. In the case of vector series that are convergent but not absolutely convergent we distinguish two cases: conditionally convergent series and unconditionally convergent series.

3.4.1 *Vector Series*

Let X be a Banach space (that is, a complete normed vector space). Let $(u_k)_{k=1}^{\infty}$ be a sequence in X and consider the series $\sum_{k=1}^{\infty} u_k$. We define the following cases, without regard at this point to any possible overlap between them:

a) The series $\sum_{k=1}^{\infty} u_k$ is *absolutely convergent* if the numerical series $\sum_{k=1}^{\infty} \|u_k\|$ is convergent.

b) The series $\sum_{k=1}^{\infty} u_k$ is *unconditionally convergent* if all its rearrangements are convergent.

c) The series $\sum_{k=1}^{\infty} u_k$ is *conditionally convergent* if it is convergent but it has a rearrangement that diverges.

Proposition 3.11 *Let X be a Banach space and $(u_k)_{k=1}^{\infty}$ a sequence in X.*

1) The series $\sum_{k=1}^{\infty} u_k$ is convergent if and only if it satisfies the following condition: for all $\varepsilon > 0$ there exists N, such that, for all $n \geq N$ and all p we have $\| \sum_{k=n}^{n+p} u_k \| < \varepsilon$.

2) The series $\sum_{k=1}^{\infty} u_k$ is unconditionally convergent if and only if it satisfies the following condition: for all $\varepsilon > 0$ there exists N, such that, for every finite set $J \subset \{N, N+1, N+2, \ldots\}$ we have $\| \sum_{k \in J} u_k \| < \varepsilon$.

Proof

1) Since $\sum_{k=n}^{n+p} u_k = w_{n+p} - w_{n-1}$, where $w_n = \sum_{k=1}^{n} u_k$, this follows immediately from the completeness of X.

2) We first assume that the condition is satisfied, let $\phi : \mathbb{N}_+ \to \mathbb{N}_+$ be a bijection and show that the rearrangement $\sum_{k=1}^{\infty} u_{\phi(k)}$ is convergent.

Let $\varepsilon > 0$. By assumption there exists N, such that for every finite set $J \subset \{N, N+1, N+2, \ldots\}$ we have $\| \sum_{k \in J} u_k \| < \varepsilon$. There also exists N_1, such that

$$\{1, \ldots, N-1\} \subset \{\phi(1), \ldots, \phi(N_1 - 1)\}.$$

Let $n \geq N_1$. For each p we have

$$\{\phi(n), \ldots \phi(n+p)\} \subset \{N, N+1, N+2, \ldots\}$$

and therefore $\| \sum_{k=n}^{n+p} u_{\phi(k)} \| < \varepsilon$. Hence the series $\sum_{k=1}^{\infty} u_{\phi(k)}$ is convergent.

Conversely, suppose that the condition in part 2 is not satisfied. We shall produce a rearrangement that diverges.

Since the condition is not satisfied, there exists $\varepsilon > 0$, such that for each N there exists a finite set $J_N \subset \{N, N+1, \ldots\}$, such that $\| \sum_{k \in J_N} u_k \| \geq \varepsilon$. We can produce an increasing sequence of natural numbers, $(K_n)_{n=1}^{\infty}$, such that the sets J_{K_n} are pairwise disjoint. For example we can proceed recursively and let $K_{n+1} = 1 + \max J_{K_n}$. Then we construct a bijection $\phi : \mathbb{N}_+ \to \mathbb{N}_+$, in such a way that it maps each set J_{K_n} to a sequence of consecutive natural numbers. The rearrangement $\sum_{k=1}^{\infty} u_{\phi^{-1}(k)}$ does not satisfy the condition in part 1 and so it is not convergent. □

The reader may have noticed that the axiom of choice was tacitly used in the proof of part 2, in order to produce the set J_N for each N. Or was it? It is risky to assert that something is "tacitly used". The reader may like to speculate as to whether the assignment of J_N to N can be defined without using the axiom of choice.

One can provide more detail for the construction of ϕ, which is, of course, highly non-unique. The suggestion made in the proof, which was to let $K_{n+1} = 1 + \max J_{K_n}$, simplifies things because it implies that

$$J_{K_n} \subset [[K_n, K_{n+1}]]$$

for each n. The sets J_{K_n} are included in consecutive natural number intervals. Then one can define ϕ so that it permutes each interval $[[K_n, K_{n+1}]]$ internally and maps J_{K_n} to a consecutive sequence. The reader should supply a constructive version of such a mapping.

Proposition 3.12 *If the series $\sum_{k=1}^{\infty} u_k$ is unconditionally convergent then all its rearrangements have the same sum.*

Proof Let $\sum_{k=1}^{\infty} u_k = w$ and assume that the series is unconditionally convergent. Consider the rearranged series $\sum_{k=1}^{\infty} u_{\phi(k)}$ and let its sum be z. We are going to show that $w = z$.

Let $\varepsilon > 0$. There exists N, such that $\| \sum_{k \in J} u_k \| < \varepsilon$ for every finite subset $J \subset \{N, N+1, N+2, \ldots\}$. Hence we also have

$$\left\| w - \sum_{k=1}^{N-1} u_k \right\| \leq \varepsilon.$$

There exists N_1, such that

$$\{1, \ldots, N-1\} \subset \{\phi(1), \ldots, \phi(N_1 - 1)\}.$$

Hence

$$\left\| \sum_{k=1}^{n} u_{\phi(k)} - \sum_{k=1}^{N-1} u_k \right\| < \varepsilon$$

for all $n \geq N_1$, since all terms u_k with $k < N$ cancel out on subtraction, and therefore also

$$\left\| w - \sum_{k=1}^{n} u_{\phi(k)} \right\| < 2\varepsilon.$$

Let $n \to \infty$. We obtain $\| w - z \| \leq 2\varepsilon$. We conclude, since ε is arbitrary, that $w = z$.   □

Proposition 3.13 *All absolutely convergent series are unconditionally convergent and all rearrangements of them have the same sum.*

Proof Suppose that $\sum_{k=1}^{\infty} \|u_k\|$ is convergent. Let $\varepsilon > 0$. There exists N, such that for all $n \geq N$ and all p we have $\sum_{k=n}^{n+p} \|u_k\| < \varepsilon$. Now if J is a finite subset of $\{N, N+1, \ldots\}$ there exists p such that $J \subset [[N, N+p]]$ and so

$$\left\| \sum_{k \in J} u_k \right\| \leq \sum_{k \in J} \|u_k\| \leq \sum_{k=N}^{N+p} \|u_k\| < \varepsilon.$$

Hence $\sum_{k=1}^{\infty} u_k$ is unconditionally convergent. By proposition 3.12 all its rearrangements have the same sum.   □

Exercise Find an example of an unconditionally convergent series that is not absolutely convergent.

Hint. Look at the spaces ℓ^p (with $p \neq \infty$) and c_0.

Proposition 3.14 *If* $X = \mathbb{R}^n$ *with the norm* $\|u\|_p$, *then every unconditionally convergent series is absolutely convergent.*

Proof Assume that the series $\sum_{k=1}^{\infty} u^{(k)}$, with $u^{(k)} \in \mathbb{R}^n$, is unconditionally convergent. Let $u^{(k)} = (u_1^{(k)}, \ldots, u_n^{(k)})$. Projecting to coordinates we see that each numerical series $\sum_{k=1}^{\infty} u_j^{(k)}$, $(j = 1, \ldots, n)$, is unconditionally convergent, and so $\sum_{k=1}^{\infty} |u_j^{(k)}| < \infty$ by Riemann's rearrangement theorem. We conclude that $\sum_{k=1}^{\infty} \|u^{(k)}\|_p < \infty$.   □

This result holds for any finite-dimensional Banach space and not just $\mathbb{R}^n$. The reason is that a finite-dimensional Banach space with dimension n is linearly homeomorphic to $\mathbb{R}^n$. This will be seen in chapter 4. The distinction between absolute convergence and unconditional convergence is therefore only important for infinite-dimensional Banach spaces.

3.4.2 Exercises

1. Two further properties of vector series that are occasionally encountered, ostensibly narrowing the notion of convergence, are:

    i) For all sequences $(\sigma_k)_{k=1}^{\infty}$, with $\sigma_k \in \{-1, 1\}$, the series $\sum_{k=1}^{\infty} \sigma_k u_k$ is convergent.
    ii) Every subseries $\sum_{k=1}^{\infty} u_{n_k}$, defined by an increasing sequence of integers $(n_k)_{k=1}^{\infty}$, is convergent.

    Show that they are both equivalent to unconditional convergence.

3.4.3 Pointers to Further Study

→   Orthonormal series.
→   Schauder bases.

3.5   (◊) Invertible Operators

We recall (section 2.4) that if X and Y are normed spaces then the space $L(X, Y)$ of continuous linear mappings $T : X \to Y$ can be equipped with a norm

$$\|T\| = \sup_{\|x\| \leq 1} \|T(x)\|.$$

We also recall from 2.4 that $\|T\|$ is the lowest constant K, such that $\|T(x)\| \leq K\|x\|$ for all $x \in X$; and that $\|T\|$ is the supremum of $\|T(x)\|$ taken over all vectors x with $\|x\| \leq 1$.

The space $L(X, Y)$ inherits completeness from Y.

Proposition 3.15 *If Y is complete then so is $L(X, Y)$.*

Proof Assume that Y is complete and let $(T_n)_{n=1}^{\infty}$ be a Cauchy sequence in $L(X, Y)$. Recall that a Cauchy sequence is bounded; hence there exists K such that $\|T_n\| \leq K$ for all n. This implies that $\|T_n(x)\| \leq K\|x\|$ for all n and all $x \in X$.

Let $x \in X$. Since

$$\|T_m(x) - T_n(x)\| \leq \|T_m - T_n\| \, \|x\|$$

it follows that $(T_n(x))_{n=1}^{\infty}$ is a Cauchy sequence in Y, hence convergent. Let its limit be $S(x)$. It is trivial to show that S is linear (it is of course a mapping from X to Y by its definition). Moreover

$$\|S(x)\| = \|\lim_{k \to \infty} T_k(x)\| = \lim_{k \to \infty} \|T_k(x)\| \leq K\|x\|.$$

It follows that S is continuous, hence an element of $L(X, Y)$.

Finally we show that T_n converges to S in the norm of $L(X, Y)$ and not just pointwise. Let $\varepsilon > 0$. There exists N, such that $\|T_m - T_n\| < \varepsilon$ for all m and n that satisfy $N \leq m < n$. In addition to this, if $x \in X$ then

$$\|T_m(x) - T_n(x)\| \leq \varepsilon\|x\|.$$

Letting $n \to \infty$ we deduce that for $m \geq N$ and all $x \in X$ we have

$$\|(T_m - S)(x)\| = \|T_m(x) - S(x)\| \leq \varepsilon\|x\|.$$

Therefore

$$\|T_m - S\| \leq \varepsilon$$

for all $m \geq N$. That is, $\lim_{m \to \infty} T_m = S$ in $L(X, Y)$.   □

Let X and Y be normed spaces. An element of $L(X, Y)$ that is bijective, and whose inverse is continuous, is called an invertible operator. Other names for it include 'linear homeomorphism'. Note that if $T \in L(X, Y)$ and T is bijective, then its inverse mapping, though rather obviously linear, is not obviously continuous. A famous theorem of functional analysis, the Banach isomorphism theorem, asserts that if X and Y are complete (they are Banach spaces to use the usual terminology) and if $T \in L(X, Y)$ is bijective, then its inverse is continuous.

Now we limit ourselves to the case that $X = Y$. We suppose that X is a Banach space and we study the space $L(X, X)$. This space is an algebra. Composition of linear mappings becomes a non-commutative product in $L(X, X)$. We recall from 2.4 that for S and T in $L(X, X)$ we have the inequality

$$\|ST\| \le \|S\| \, \|T\|.$$

The invertible elements of $L(X, X)$ form a group $GL(X)$. If X is finite dimensional we can study this using matrices and in effect we have here the general linear group $GL(n)$, of invertible $n \times n$-matrices. Since the invertible matrices are those with non-zero determinant, and since the determinant is a continuous function of the n^2 entries of the matrix (it is simply a polynomial with n^2 variables), it follows that if X is finite dimensional the space $GL(X)$ is an open subset of $L(X, X)$. However, if X is infinite dimensional the determinant is not available. Nevertheless we can still show that $GL(X)$ is an open subset of $L(X, X)$. It is an interesting application of absolutely convergent series in a Banach space.

Proposition 3.16 *Let X be a Banach space. Let $T \in L(X, X)$ and suppose that $\|T\| < 1$. Then the operator $I + T$ is invertible and its inverse is given by the series*[1]

$$(I + T)^{-1} = \sum_{n=0}^{\infty} (-1)^n T^n. \tag{3.4}$$

The series is absolutely convergent in the space $L(X, X)$.

Proof We have that

$$\|(-1)^n T^n\| \le \|T\|^n.$$

---

[1] Sometimes called the Neumann series, after C. Neumann, a 19th century mathematician who should not be confused with John von Neumann.

Since the space $L(X, X)$ is complete and $\|T\| < 1$ the series $\sum_{n=0}^{\infty} T^n$ is absolutely convergent in $L(X, X)$. Moreover,

$$(I + T) \sum_{n=0}^{\infty} (-1)^n T^n = \sum_{n=0}^{\infty} (-1)^n T^n + \sum_{n=0}^{\infty} (-1)^n T^{n+1}$$

$$= \sum_{n=0}^{\infty} (-1)^n T^n + \sum_{n=1}^{\infty} (-1)^{n-1} T^n$$

$$= I + \sum_{n=1}^{\infty} \left((-1)^n + (-1)^{n-1} \right) T^n = I.$$

If one prefers, the above calculation can be justified by taking a finite sum and going to the limit. To conclude the proof of (3.4) we must look at the reversed product. Virtually the same calculation gives

$$\left(\sum_{n=0}^{\infty} (-1)^n T^n \right) (I + T) = I.$$

□

Exercise Obtain the useful estimate

$$\|(I + T)^{-1}\| \leq \frac{1}{1 - \|T\|}$$

under the condition $\|T\| < 1$.

Proposition 3.17 *Let X be a Banach space. Then the group $GL(X)$ of invertible operators is open in $L(X, X)$.*

Proof Let $A \in GL(X)$. Then for $B \in L(X, X)$ we have

$$A + B = A(I + A^{-1}B).$$

Now if $\|B\| < \|A^{-1}\|^{-1}$ we have

$$\|A^{-1}B\| \leq \|A^{-1}\| \|B\| < 1.$$

so that $I + A^{-1}B$ is invertible and therefore also $A + B$. Putting it differently: if A is invertible then so is T whenever $\|T - A\| < \|A^{-1}\|^{-1}$.   □

Exercise Obtain the estimate

$$\|(A + B)^{-1}\| \leq \frac{\|A^{-1}\|}{1 - \|A^{-1}\| \|B\|}.$$

under the condition that $\|B\| < \|A^{-1}\|^{-1}$.

3.5.1 Fredholm Integral Equation

We will give an application of proposition 3.16 to solving an integral equation. Let X be the Banach space $C[a, b]$ with its usual norm $\|u\|_\infty = \sup_{a \le x \le b} |u(x)|$, and let T be the element of $L(X, X)$ defined by

$$(Tf)(x) = \int_a^b K(x, y) f(y) \, dy.$$

The function $K(x, y)$, called the kernel of the transformation T, is subject to the following conditions:

1. $K(x, y)$ is continuous in the square $[a, b] \times [a, b]$.
2. If M is the maximum of $|K(x, y)|$ in $[a, b] \times [a, b]$ then $M(b - a) < 1$.

Exercise Show that Tf really is continuous in $[a, b]$ if f is continuous in $[a, b]$, as was implicit above in saying that T was an element of $L(X, X)$. You will need a small acquaintance with the analysis of continuous functions of two real variables.

Given the function $h(x)$, continuous for $a \le x \le b$, we are going to solve the equation

$$u(x) - \int_a^b K(x, y) u(y) \, dy = h(x), \quad (a \le x \le b). \tag{3.5}$$

The unknown quantity sought here is the function $u(x)$.

Under the given conditions we find, for a given $u \in X$, that

$$|Tu(x)| \le M(b - a)\|u\|_\infty, \quad (a \le x \le b)$$

and so

$$\|Tu\|_\infty \le M(b - a)\|u\|_\infty$$

This implies, for the operator norm, that $\|T\| \le M(b - a) < 1$. The solution is therefore given by the series

$$u = \left(\sum_{n=0}^\infty T^n \right) h = \sum_{n=0}^\infty T^n h.$$

The convergence of the second series is in $C[a, b]$, that is, we have uniform convergence for $a \le x \le b$.

Exercise Show that the operator T^n may also be expressed using a kernel K_n, where $K_1 = K$ and

$$K_{n+1}(x, y) = \int_a^b K_n(x, t) K(t, y) \, dt.$$

This can be viewed as an n-fold integral

$$K_{n+1}(x, y) = \int_a^b \cdots \int_a^b K(x, t_1) K(t_1, t_2) \ldots K(t_{n-1}, t_n) K(t_n, y) \, dt_1 \ldots dt_n.$$

You will need Fubini's theorem for Riemann-Darboux integrals of two real variables, usually covered in a first course of multivariate calculus. Now consider the function

$$R(x, y) = \sum_{n=1}^{\infty} K_n(x, y), \quad (a \le x, y \le b).$$

Show that the series is uniformly convergent in the square $[a, b] \times [a, b]$. Deduce from this that the function $R(x, y)$ is continuous and that the solution of (3.5) may be written

$$u(x) = h(x) + \int_a^b R(x, y) h(y) \, dy, \quad (a \le x \le b).$$

3.5.2 Exercises

1. Let X be a Banach space and let $T \in L(X, X)$. Show that if the series $\sum_{n=0}^{\infty} (-1)^n T^n$ converges then its sum is the inverse of $I + T$.
2. In the previous exercise the norm of T could be bigger than 1. Give an example of an operator T with norm bigger than 1 for which the series converges.
   Hint. Find examples of $T \in L(\mathbb{R}^2, \mathbb{R}^2)$ for which $T^2 = 0$ and $\|T\|$ is arbitrarily large.
3. An important case when convergence of the series (3.4) can be inferred without a bound on $\|T\|$ occurs in connection with the Volterra integral equation. This is a problem of the form

$$u(x) - \int_a^x K(x, y) u(y) \, dy = h(x), \quad (a \le x \le b) \tag{3.6}$$

where the unkown function $u \in C[a, b]$ is sought given the function $h \in C[a, b]$. The kernel function $K(x, y)$ is defined and continuous in the triangle

$$D := \{(x, y) : a \le y \le x \le b\}.$$

No condition is needed on the maximum M of $K(x, y)$ in D.

Letting $X = C[a, b]$ we define the operator $T \in L(X, X)$ by

$$(Tu)(x) = \int_a^x K(x, y)u(y)\, dy, \quad (a \le x \le b,\ u \in X).$$

In the following steps you will need some easy bivariate calculus and Fubini's theorem.

a) Show that T really does map X into X and is linear.
b) Show that for each n we have

$$(T^n u)(x) = \int_a^x K_n(x, y)u(y)\, dy, \quad (a \le x \le b,\ u \in X)$$

where K_n satisfies the recurrence relation

$$K_{n+1}(x, y) = \int_y^x K(x, t)K_n(t, y)\, dt, \quad (a \le y \le x \le b)$$

c) Derive the estimate

$$|K_n(x, y)| \le \frac{M^n(x - y)^{n-1}}{(n - 1)!}, \quad (a \le y \le x \le b,\ n = 1, 2, 3, \ldots)$$

d) Derive the estimate

$$\|T^n\| \le \frac{M^n(b - a)^n}{n!}, \quad (n = 1, 2, 3, \ldots)$$

e) Deduce that the series $\sum_{n=0}^{\infty} T^n$ converges in $L(X, X)$ and infer that the problem (3.6) has a unique solution.

f) Show that the solution of (3.6) can be expressed in the form

$$u(x) = h(x) + \int_a^x L(x, y)h(y)\, dy \quad (a \le x \le b)$$

for a suitable kernel $L(x, y)$ continuous in the triangle D.

4. Let X be a Banach space and let A be an invertible element of $L(X, X)$. We have seen that if $T \in L(X, X)$ and $\|T\|\,\|A^{-1}\| < 1$ then $A + T$ is invertible. Show

that, also, we have the estimate

$$\|(A + T)^{-1} - A^{-1} + A^{-1}T A^{-1}\| \leq \frac{\|A^{-1}\|^3 \|T\|^2}{1 - \|T\|\,\|A^{-1}\|}$$

Note. For readers studying Fréchet derivatives. The inequality says that the mapping $A \mapsto A^{-1}$ has the derivative $T \mapsto -A^{-1}T A^{-1}$ at A.

3.5.3 Pointers to Further Study

$\to$ Functional analysis.
$\to$ Operator algebras.
$\to$ Integral equations.

3.6 ($\Diamond$) Tietze

This section is devoted to one problem. A continuous function is defined on a subset of a metric space. We want to extend it to a continuous function defined on the whole space. This is the content of Tietze's extension theorem. Many proofs of this famous theorem are known. We are going to use uniform convergence of function series and the Weierstrass M-test.

Proposition 3.18 (Tietze's Extension Theorem) *Let X be a metric space, let $V \subset X$ be closed and let $g : V \to \mathbb{R}$ be continuous. Then there exists a continuous function $f : X \to \mathbb{R}$, such that $f|_V = g$. If, moreover, there exist a and b, such that $a \leq g(x) \leq b$ for all $x \in V$, then the extension f can be built to satisfy $a \leq f(x) \leq b$ for all $x \in X$.*

The main prerequisite for the proof is another famous result, actually a special case of the Tietze extension theorem.

Proposition 3.19 (Urysohn's Lemma) *Let X be a metric space and let A and B be disjoint closed sets in X. Then there exists a continuous function $f : X \to \mathbb{R}$, such that:*

1) For all $x \in X$ we have $0 \leq f(x) \leq 1$.
2) For all $x \in A$ we have $f(x) = 0$.
3) For all $x \in B$ we have $f(x) = 1$.

Proof If $A \neq \emptyset$ and $B \neq \emptyset$ we set

$$f(x) = \frac{d(x, A)}{d(x, A) + d(x, B)}.$$

If $A = \emptyset$ but $B \neq \emptyset$ we set $f = 1$. If $B = \emptyset$ but $A \neq \emptyset$ we set $f = 0$. If both are empty we set $f = 0$. □

Given constants a and b, by composing the function f delivered by the last proposition with a first degree polynomial, we can build a function g that satisfies $a \leq g \leq b$, $g|_A = a$, $g|_B = b$. This will be used repeatedly in what follows.

Lemma 3.1 *Let X be a metric space and let V be a closed set in X. Let $g : V \to \mathbb{R}$ be a continuous function and assume that there exists $c > 0$, such that $|g(x)| \leq c$ for all $x \in V$. Then there exists a continuous function $f : X \to \mathbb{R}$ with the following properties:*

1) $|f(x)| \leq \frac{1}{3}c$, $(x \in X)$.
2) $|g(x) - f(x)| \leq \frac{2}{3}c$, $(x \in V)$.

Proof Let

$$A = \{x \in V : g(x) \geq \tfrac{1}{3}c\}$$
$$B = \{x \in V : g(x) \leq -\tfrac{1}{3}c\}.$$

The sets A and B are closed relative to V, but since V is itself closed in X, they are also closed in X. We use proposition 3.19 (actually the remark following it) to build a continuous function $f : X \to \mathbb{R}$ that satisfies:

$$-\tfrac{1}{3}c \leq f(x) \leq \tfrac{1}{3}c, \quad (x \in X),$$

$$f(x) = \begin{cases} \tfrac{1}{3}c, & (x \in A), \\ -\tfrac{1}{3}c, & (x \in B). \end{cases}$$

Obviously f has the sought-after properties (illustrated in Fig. 3.1). □

Proof of Proposition 3.18 We shall consider the case when $|g(x)| \leq c$ for $x \in V$ and build an extension that satisfies the same inequalities in X. The case when $a \leq g(x) \leq b$ for $x \in V$, is obtained from this by composing g with a first degree polynomial.

By the lemma there exists a continuous function $f_1 : X \to \mathbb{R}$ that satisfies the two conditions:

$$|f_1(x)| \leq \tfrac{1}{3}c, \quad (x \in X),$$
$$|g(x) - f_1(x)| \leq \tfrac{2}{3}c, \quad (x \in V).$$

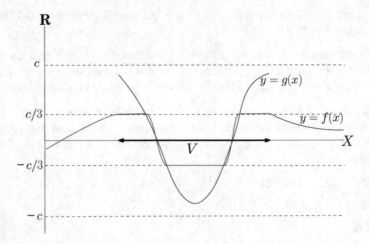

Fig. 3.1 Picture of the lemma

We repeat the step with $g - f_1$ instead of g and $\frac{2}{3}c$ instead of c. We build f_2 that satisfies the two conditions:

$$|f_2(x)| \le \tfrac{2}{9}c, \quad (x \in X),$$

$$|g(x) - f_1(x) - f_2(x)| \le \tfrac{4}{9}c, \quad (x \in V).$$

Continuing by induction, we build a sequence $(f_n)_{n=1}^{\infty}$ of functions that satisfy the two conditions:

$$|f_n(x)| \le \frac{2^{n-1}c}{3^n}, \quad (x \in X), \tag{3.7}$$

$$\left| g(x) - \sum_{j=1}^{n} f_k(x) \right| \le \frac{2^n c}{3^n}, \quad (x \in V). \tag{3.8}$$

By (3.7) and the Weierstrass M-test the series $\sum_{n=1}^{\infty} f_n$ is uniformly convergent in X and its sum f is a continuous function in X. By (3.8) we see that $g = f$ in V and by (3.7) that

$$|f| \le \sum_{n=1}^{\infty} |f_n| \le \sum_{n=1}^{\infty} \frac{2^{n-1}c}{3^n} = c.$$

If g is unbounded we need a new argument. We let $h = \arctan \circ g$. By extending h we obtain a suitable extension of g. There are some details that the reader is asked to provide (see exercise 2). $\qquad \square$

3.6.1 Formulas for an Extension

Tietze's extension theorem can be proved by writing down a formula for an extension. We give two examples here without proofs. The main challenge is to show that the formula defines a continuous function.

(A) (Hausdorff's formula)

$$
f(x) = \begin{cases} \inf_{y \in V} \left(g(y) + \dfrac{d(x, y)}{d(x, V)} - 1 \right), & (x \notin V) \\ g(x), & (x \in V) \end{cases}
$$

(B) (Dieudonné, Foundations of Modern Analysis)

Given that $1 \le g \le 2$ in V the following extension satisfies $1 \le f \le 2$:

$$
f(x) = \begin{cases} \dfrac{\inf_{y \in V} \left(g(y) d(x, y) \right)}{d(x, V)}, & (x \notin V) \\ g(x), & (x \in V) \end{cases}
$$

3.6.2 Exercises

1. Let X be a metric space and let A and B be disjoint closed sets in X. Exercise 25 in 2.2 asked for a proof that there exist disjoint open sets U and V, such that $A \subset U$ and $B \subset V$. Show that this also follows from Urysohn's lemma.
2. The case of proposition 3.18 when g is unbounded is treated by letting $h(x) = \arctan(g(x))$ for $x \in V$. Now $-\pi/2 < h(x) < \pi/2$ for all $x \in V$ and we can extend h to a continuous function $\tilde{h}$ on X. However, we can only guarantee that $-\pi/2 \le \tilde{h}(x) \le \pi/2$, and since we want to define the extension f of g by $f(x) := \tan(\tilde{h}(x))$ we have a problem if there are points x such that $\tilde{h}(x) = \pm\pi/2$. Overcome this problem in two easy steps:

   a) Let W be the set of points in X at which $\tilde{h}(x) = \pm\pi/2$. Show that W is closed and that $W \cap V = \emptyset$.
   b) Let ϕ be a continuous function on X satisfying $0 \le \phi \le 1$, $\phi(x) = 1$ for $x \in V$ and $\phi(x) = 0$ for $x \in W$. Define

$$
f(x) := \tan(\phi(x)\tilde{h}(x)), \quad (x \in X).
$$

   Show that f is the sought-for extension of g.

3. Let X be a metric space. The support of a function $f : X \to \mathbb{R}$ (see 2.3 exercise 16) is the set

$$\text{supp}(f) := \overline{\{x \in X : f(x) \neq 0\}}.$$

Let M be a closed set, and U an open set such that $M \subset U$.

a) Show that there exists a continuous function $f : X \to [0, 1]$, such that $f = 1$ in M but $\text{supp}(f) \subset U$.
Hint. Use Urysohn's lemma.

b) Suppose that $g : M \to [a, b]$ is continuous. Show that g has an extension $h : X \to [a, b]$ such that $\text{supp}(h) \subset U$.

4. Let X be a metric space and let $(U_k)_{k=1}^{n}$ be a sequence of open sets such that

$$X = \bigcup_{k=1}^{n} U_k.$$

Show that there exist continuous functions $f_k : X \to [0, 1]$, $(k = 1, \ldots, n)$, such that

$$\text{supp}(f_k) \subset U_k, \quad (k = 1, \ldots, n)$$

and

$$\sum_{k=1}^{n} f_k = 1.$$

For a hint see after the following exercise.

5. Extend the result of the previous exercise to the following case: M is closed in X and $(U_k)_{k=1}^{n}$ is a sequence of open sets such that

$$M \subset \bigcup_{k=1}^{n} U_k.$$

Prove that there exist continuous functions $f_k : X \to [0, 1]$, $(k = 1, \ldots, n)$, such that

$$\text{supp}(f_k) \subset U_k, \quad (k = 1, \ldots, n)$$

$$\sum_{k=1}^{n} f_k(x) \leq 1, \quad (x \in X).$$

and

$$\sum_{k=1}^{n} f_k(x) = 1, \quad (x \in M).$$

Hint for exercise 4. Apply 2.2 exercise 27 to find open sets V_k that satisfy $V_k \subset \overline{V_k} \subset U_k$ and $X = \bigcup_{k=1}^{n} V_k$. Apply the previous exercise to find continuous functions $g_k : X \to [0, 1]$ equal to 1 in $\overline{V_k}$ with support in U_k. Now let $f_k = g_k / \sum_{i=1}^{n} g_i$.

Hint for exercise 5. Find open sets V_k such that $V_k \subset \overline{V_k} \subset U_k$ and $M \subset \bigcup_{k=1}^{n} V_k$. Find continuous functions $g_k : X \to [0, 1]$ equal to 1 in $\overline{V_k}$ with support in U_k. The definition $f_k = g_k / \sum_{i=1}^{n} g_i$ now doesn't work because the denominator can have zeros. There are two approaches and the reader is invited to choose the one they prefer (or find another).

Approach 1. Let $m = \inf_M \sum_{i=1}^{n} g_i$, note that $m \geq 1$, and find a continuous function $\chi : X \to [0, 1]$ equal to 1 where $\sum_{i=1}^{n} g_i \geq m$ and equal to 0 where $\sum_{i=1}^{n} g_i \leq \frac{1}{2}m$. Make sense of the "definition" $f_k = \chi g_k / \sum_{i=1}^{n} g_i$.

Approach 2. Let

$$f_k = (1 - g_1) \ldots (1 - g_{k-1}) g_k, \quad (k = 1, \ldots, n)$$

Note. Exercises 4 and 5 introduce simple examples of *partitions of unity*. They are capable of generalisation in several ways. One is to admit infinitely many open sets. Another is to require the functions to be differentiable, in a context where differentiability makes sense (for example on a differentiable manifold). A third is to replace M by a *compact set* and to require the supports to be compact sets. Compact sets will be studied in the next chapter. Partitions of unity provide a valuable tool in higher calculus.

3.6.3 *Pointers to Further Study*

→   General topology.
→   Extension theorems.

Chapter 4
Compact Spaces

The lunatic, the lover and the poet
Are of imagination all compact

Shakespeare. A Midsummer Night's Dream

Does the space $[a, b]$ possess a topological property that implies that all continuous functions on $[a, b]$ are bounded; and attain their maximum and minimum? At first sight this question seems fatuous, since the property that every continuous function $f : X \to \mathbb{R}$ is bounded is itself a topological property of the metric space X. We are after something else here. We wish to peer into the structure of $[a, b]$ to find a property that implies this, and which will be applicable to a wide range of metric spaces.

4.1 Sequentially Compact Spaces

By the Bolzano-Weierstrass theorem every bounded sequence in $\mathbb{R}$ has a convergent subsequence. Hence every sequence in $[a, b]$ has a convergent subsequence, and the limit of the subsequence is in $[a, b]$ because the latter is a closed set in $\mathbb{R}$. If we view $[a, b]$ as a metric space in its own right, a subspace of $\mathbb{R}$, then we can say that every sequence in $[a, b]$ has a convergent subsequence. We referred to this property in 2.5; now we define it formally.

Definition A metric space (M, d) is said to be *sequentially compact* if every sequence in M has a convergent subsequence.

This is clearly a topological property, since convergence of a sequence can be defined using open sets without mentioning the metric.

The property of sequential compactness applies to spaces. But we can define what it means for a *set in a metric space* to be sequentially compact. This uses a common device, whereby a notion applicable to spaces is extended in scope, so that it can be applied to sets in spaces. This should eliminate any confusion between a sequentially compact space and a sequentially compact set in a space.

© The Author(s), under exclusive license to Springer Nature Switzerland AG 2022
R. Magnus, *Metric Spaces*, Springer Undergraduate Mathematics Series,
https://doi.org/10.1007/978-3-030-94946-4_4

Definition Let (M, d) be a metric space. A subset $A \subset M$ is called a sequentially compact set in M if the space (A, d) is a sequentially compact space.

It should be obvious that A is a sequentially compact set if and only if every sequence in A has a convergence subsequence with limit in A.

4.1.1 Continuous Functions on Sequentially Compact Spaces

The question asked at the beginning of the chapter is easily answered.

Proposition 4.1 Let (M, d) be a sequentially compact metric space and let f : $M \to \mathbb{R}$ be continuous. Then f is bounded (that is, there exists K, such that $|f(x)| < K$ for all $x \in M$). Moreover f attains its maximum and minimum.

Proof We assume that f is not bounded above and derive a contradiction. There exists a sequence $(x_n)_{n=1}^{\infty}$ in M, such that $f(x_n) \to \infty$. Let $(x_{k_n})_{n=1}^{\infty}$ be a convergent subsequence and let its limit be $a \in M$. Then $f(x_{k_n}) \to f(a)$ (because f is continuous), but also $f(x_{k_n}) \to \infty$, which is a contradiction. A similar argument shows that f is bounded below.

We show next that f attains its supremum and infimum. Let $p = \sup_{x \in M} f(x)$ and $q = \inf_{x \in M} f(x)$. There exists a sequence $(x_n)_{n=1}^{\infty}$ in M, such that $f(x_n) \to p$. Let $(x_{k_n})_{n=1}^{\infty}$ be a convergent subsequence and let its limit be $a \in M$. Then $f(a) = \lim_{n \to \infty} f(x_{k_n}) = p$. Similarly there exists b, such that $f(b) = q$. □

The axiom of choice was needed several times in this proof. Where? For more on this see 5.2 exercise 15.

4.1.2 Bolzano-Weierstrass in $\mathbb{R}^n$

We come to an indispensable result for multivariate calculus. Its applications are legion. Nowadays we are apt to take it for granted, but it should be noted that it depends crucially on the completeness property of the real numbers, a concept that was slow to be clarified. An truly exact proof of the fundamental theorem of algebra eluded Gauss because he lacked it.

Proposition 4.2 (Bolzano-Weierstrass) Equip $\mathbb{R}^n$ with any of the norms $\|x\|_p$ $(1 \le p \le \infty)$. Then every bounded sequence in $\mathbb{R}^n$ has a convergent subsequence.

Proof Let $(x^{(i)})_{i=1}^{\infty}$ be a sequence in $\mathbb{R}^n$, bounded with respect to the norm $\|x\|_p$. Then, for each $k = 1, \ldots, n$, the numerical sequence of the kth coordinates, $(x_k^{(i)})_{i=1}^{\infty}$, is bounded. Choose a subsequence of the vector sequence $(x^{(i)})_{i=1}^{\infty}$, such that its sequence of first coordinates is convergent in $\mathbb{R}$. Next we choose a subsequence of the first subsequence, such that the sequence of second coordinates is convergent. Then a subsequence of the second subsequence, and so on. After a finite number of steps we reach a subsequence of the initial sequence, such that for each of $k = 1, 2, \ldots, n$, the sequence of the kth coordinates is convergent. This final subsequence is convergent in $\mathbb{R}^n$. □

4.1.3 Sequentially Compact Sets in $\mathbb{R}^n$

We can identify all sequentially compact sets in $\mathbb{R}^n$.

Proposition 4.3 *A set $A \subset \mathbb{R}^n$ is sequentially compact if and only if it is bounded and closed.*

Proof If A is bounded and closed then every sequence in A has a convergent subsequence (by Bolzano-Weierstrass), and since A is closed the limit of the subsequence lies in A. We conclude that the set A is sequentially compact.

If A is not closed then there exists a sequence $(a^{(i)})_{i=1}^{\infty}$ in A that is convergent but its limit is not in A. This sequence does not have a convergent subsequence with limit in A, so that A is not sequentially compact.

If A is unbounded then there exists a sequence $(a^{(i)})_{i=1}^{\infty}$ in A, such that $\|a^{(i)}\|_p > 2^i$ for all i. Obviously this sequence does not have a convergent subsequence, so that A is not sequentially compact. □

The importance of this result is that, alongside proposition 4.1, it supplies a proof that solutions to optimisation problems exist in appropriate circumstances, by providing a plentiful supply of sequentially compact sets. Since optimisation is a frequently encountered problem of applied mathematics it is hard to overestimate the value of proposition 4.3.

For example, let $A \subset \mathbb{R}^3$ be the set

$$\{(x, y, z) : x \geq 0, \ y \geq 0, \ z \geq 0, \ 2x + 3y + 4z = 1\}.$$

Suppose that is required to maximise or minimise a function, for example $f(x, y, z) := xy^2 + 2yz^2 + 3zx^2$, on the set A. Propositions 4.1 and 4.3 guarantee that both maximum and minimum are attained.

Exercise Write out the arguments justifying this claim.

Practical methods for finding maxima and minima, as well as the points where they are attained, are in the domain of applied mathematics.

4.1.4 Sequentially Compact Sets in Other Spaces

The 'only if' part of proposition 4.3 is valid in every metric space and the proof is effectively the same. For a subset A of a metric space X to be compact it is necessary that A is closed and bounded. But this is far from being sufficient in general. For example it is not sufficient in ℓ^p, nor in $C[a, b]$, nor in any infinite-dimensional normed vector space.

In all infinite-dimensional normed vector spaces one can find a bounded sequence with no convergent subsequence. The reader is invited to find examples in the spaces ℓ^p and $C[a, b]$. It is an important question to seek conditions, in addition to those of being closed and bounded, that suffice for a set to be compact in such spaces. The famous Arzela-Ascoli theorem solves this question for the space $C[a, b]$.

An important example of a sequentially compact set in ℓ^2 is the *Hilbert cube*. This is the set of all sequences $x = (x_n)_{n=1}^{\infty} \in \ell^2$, such that $0 \leq x_n \leq \frac{1}{n}$ for each n. We will study this set later.

4.1.5 The Space $C(M)$

Let M be a sequentially compact metric space. Since all continuous functions $f : M \to \mathbb{R}$ are bounded we can use the norm $\|f\|_{\infty} = \sup_{x \in M} |f(x)|$ without stipulating that the functions are bounded. The proof of the following proposition is the same as that of the corresponding one for bounded functions.

Proposition 4.4 *If M is sequentially compact then $C(M)$ is complete with the metric $\|f\|_{\infty}$.*

The space $C(M)$ has some important properties. As well as being a complete normed space (that is, a Banach space) it is also an *algebra* and a *lattice*. These claims are embodied in the following proposition. The proofs are exercises.

Proposition 4.5 *Let $f, g \in C(M)$. Then*

1) $|f| \in C(M)$.
2) $\max(f, g)$ *and* $\min(f, g)$ *are in* $C(M)$.
3) $fg \in C(M)$ *and* $\|fg\|_{\infty} \leq \|f\|_{\infty}\|g\|_{\infty}$.

Item 2 of the proposition says that $C(M)$ is a *lattice*; item 3 that $C(M)$ is a *normed algebra*. An algebra is a vector space over $\mathbb{R}$ or $\mathbb{C}$ that is equipped with a multiplication enabling one to form the product of two vectors, the product being a vector in the same space. The multiplication should be bilinear, but is not necessarily commutative. Multiplicative inverses are not required. Nor is it required that the multiplication is associative, but non-associative algebras are uncommon objects of study. We list some examples:

a) $\mathbb{C}$ is a two-dimensional vector space over $\mathbb{R}$. It is an algebra (actually a field) with complex multiplication.
b) $\mathbb{R}^3$ is a 3-dimensional vector space over $\mathbb{R}$. The cross-product $u \times v$ makes it into an algebra. It is neither commutative nor associative.
c) The space of $(n \times n)$-real matrices is an n^2-dimensional vector space over $\mathbb{R}$. It is a non-commutative algebra under matrix multiplication.

d) The space $L(X, X)$ of continuous linear mappings $T : X \to X$, where X is a normed space. Composition of mappings provides the product.
e) $C(M)$ is a vector space over $\mathbb{R}$. The point-wise product of functions $(fg)(x) = f(x)g(x)$ makes it into a commutative algebra.

A lattice is an ordered set in which every pair of elements has a supremum and an infimum. The ordering of $C(M)$ is pointwise: $f \leq g$ if and only if $f(x) \leq g(x)$ for all $x \in M$. The supremum and infimum of f and g are $\max(f, g)$ and $\min(f, g)$.

4.1.6 Exercises

1. Let $f : [0, 1] \to \mathbb{R}^2$ be a continuous and injective mapping. Show that the inverse $f^{-1} : C \to [0, 1]$, where C is the range of f (the geometric curve parametrised by f), is continuous. Compare 2.5 exercise 2.

   Hint. Let $(p_n)_{n=1}^{\infty}$ be a sequence in C, converging to $q \in C$, and consider the sequence $(f^{-1}(p_n))_{n=1}^{\infty}$ in $[0, 1]$.
2. a) Let A be a sequentially compact set in a metric space X. Let $x \notin A$. Show that there exists $y \in A$, such that $d(x, y) = d(x, A)$.
   b) Show that the result of item (a) still holds if $X = \mathbb{R}^n$ and A is an *unbounded* closed set in $\mathbb{R}^n$, even though in this case A is not sequentially compact.
   c) The result of item (a) may fail if A is closed but not sequentially compact. Find an example in which X is a subspace of $\mathbb{R}$.
3. Let A be a sequentially compact set in a metric space X. We saw in the previous exercise that for each $x \in X$ there exists $y \in A$, such that $d(x, y) = d(x, A)$.

   a) Show by an example that the point y thus obtained does not have to be uniquely defined.
   b) Show that if y is uniquely defined for each $x \in X$, then the mapping $F(x) = y$, so defined, from X to A, is continuous.

4. For non-empty disjoint sets A and B in a metric space X we define their *separation* by

$$\text{sep}(A, B) = \inf\{d(x, y) : x \in A, \ y \in B\}.$$

   Find an example of a metric space X and two non-empty disjoint closed sets A and B such that $\text{sep}(A, B) = 0$. Show, however, that in general if A is sequentially compact, B closed, and A and B disjoint, then $\text{sep}(A, B) > 0$.
5. Let $(A_n)_{n=1}^{\infty}$ be a sequence of sequentially compact sets in a metric space X, such that none of them is empty and $A_{n+1} \subset A_n$ for each n. Show that $\bigcap_{n=1}^{\infty} A_n \neq \emptyset$.
6. If M is a non-compact metric space there is a useful subspace of $C(M)$ that can be equipped with a norm. This is the space $C_0(M)$, comprising real-valued continuous functions with domain M that "vanish at infinity". By definition an element of $C(M)$ is in $C_0(M)$ if it satisfies the following condition: for every

$\varepsilon > 0$ there exists a (sequentially) compact set $K \subset M$ such that $|f(x)| < \varepsilon$ for all $x \in M \setminus K$.

a) Show that $C_0(M)$ is a closed linear subspace of $C_B(M)$.
b) Show that $C_0(M)$ is a Banach space under the norm $\|f\|_\infty$.

Note. If M is compact then $C_0(M) = C(M)$ by default.

7. a) Let $f : \mathbb{R}^n \to \mathbb{R}$ be continuous and satisfy $\lim_{|x| \to \infty} f(x) = \infty$. Show that f attains a minimum in $\mathbb{R}^n$.
   b) Let $f : \mathbb{R}^n \to [0, \infty[$ be continuous and satisfy $\lim_{|x| \to \infty} f(x) = 0$. Show that f attains a maximum in $\mathbb{R}^n$.
8. Prove the Fundamental Theorem of Algebra. Every complex non-constant polynomial $P(z)$ has a root. You can use the following steps:

   a) Show that $\lim_{|z| \to \infty} |P(z)| = \infty$.
   b) Deduce by the previous exercise that $|P(z)|$ attains a minimum in $\mathbb{C}$.
   c) Let the minimum of $|P(z)|$ be attained at the point z_0. Show that if $P(z_0) \neq 0$ then, sufficiently near to z_0, there exists z, such that the absolute value $|P(z)|$ is strictly less than $|P(z_0)|$. Hence deduce that $P(z_0) = 0$.
   Hint. Show that one may write

$$P(z) = \alpha - \beta(z - z_0)^r G(z),$$

where $\alpha = P(z_0)$, $\beta \neq 0$, r is a positive integer and $G(z)$ is a polynomial such that $G(z_0) = 1$. Now let $z = z_0 + \gamma$, where γ is chosen to satisfy $\gamma^r = \alpha \beta^{-1}$, and consider $|P(z)|$ for z sufficiently near to z_0.

4.2   The Correct Definition of Compactness

Difficult questions in analysis led to the formulation of the most important concepts of metric space theory. A problem that was most influential was that of finding necessary and sufficient conditions for a bounded function on an interval $[a, b]$ in $\mathbb{R}$, or more generally on a cube $T = [a_1, b_1] \times \ldots \times [a_n, b_n]$ in $\mathbb{R}^n$, to be Riemann-integrable. Consider for example the following proposition, of basic importance in integration theory:

A continuous real-valued function f on a cube T is uniformly continuous.

Uniform continuity of f in T means that for all $\varepsilon > 0$ there exists $\delta > 0$, such that $|f(x) - f(y)| < \varepsilon$ for all x and y in T that satisfy $|x - y| < \delta$.

It is true that this can be proved by the method of bisection, and this involves calling in the Bolzano-Weierstrass theorem. This is easy in one dimension, but unwieldy in many.

An argument of a quite different character began to appear in the work of nineteenth century mathematicians. For each $x \in T$ one can find a "small" cube S_x

with x at its centre, in which the oscillation of f is small. It appeared that one could always select a finite number of these "small" cubes that sufficed to cover all of T. This was the genesis of the Heine-Borel theorem, and of a concept of compactness that has more general validity than that of sequential compactness when it comes to studying topological spaces that are not metric spaces, although for metric spaces the two notions are equivalent. They could hardly be more different in character.

Definition Let (X, d) be a metric space. An open covering of X is a family $(U_i)_{i \in I}$ of sets, each of which is an open set in X, such that $X = \bigcup_{i \in I} U_i$.

We say loosely "the open sets U_i cover X", when $(U_i)_{i \in I}$ is an open covering of X.

Definition The metric space (X, d) is said to be compact if every open covering of X has a finite subcovering.

We can phrase this as follows: whenever the family $(U_i)_{i \in I}$ of open sets covers X, there exists a finite subset $\{i_1, \ldots, i_n\} \subset I$, such that the family $(U_{i_k})_{k=1}^{n}$ covers X.

It is clear that compactness is a topological property. It is defined only with reference to the system of open sets in X. For a while we will take an exploratory attitude and develop properties of compact spaces, only later will we show that compactness and sequential compactness are equivalent.

4.2.1 Thoughts About the Definition

The definition of compact space has the form '$(\forall D)(\ldots)$'. In this respect it resembles the definition of limit of a sequence where the opening existential quantifier ranges over all real numbers. This time the quantifier ranges over a set of more sophisticated objects: the set of all open coverings of X. Essentially we quantify over all subsets of the power set $\mathcal{P}(X)$, that is, over the set $\mathcal{P}(\mathcal{P}(X))$, with a restriction to those comprising open sets and whose union is X.

In order to apply the definition to prove that a space is compact we must say "Let $(U_i)_{i \in I}$ be an open covering of X"; and show that there exists a finite set $\{i_1, \ldots, i_n\} \subset I$, such that $X = \bigcup_{k=1}^{n} U_{i_k}$, making it clear that no additional property of the initial open covering is needed for the purpose. However, the definition is often applied to prove that a space is compact, by contradiction. Therefore it is essential to be clear about how to deny that a space is compact. One way to phrase it is as follows:

> There exists a covering $(U_i)_{i \in I}$ of X by open sets, such that, for all finite sets $\{i_1, \ldots, i_n\} \subset I$, there exists $x \in X$, such that $x \notin \bigcup_{k=1}^{n} U_{i_k}$.

We opened the chapter by asking what property of a space X ensures that all continuous functions $f : X \to \mathbb{R}$ are bounded. Suppose we ask the same question about sets and functions—no metric, no question of continuity. What property of a set X will ensure that all functions $f : X \to \mathbb{R}$ are bounded? It is an easy exercise for the reader to show that the required property is finiteness. Therefore it is tempting to answer the corresponding question for metric spaces and continuous functions by seeking a kind of "topological finiteness". The concept of compactness accomplishes this by incorporating in a direct way both the notion of finiteness and that of open set. We really seem to peer into the structure of the space, as we put it in the preamble to this chapter.

It is instructive to try to construct open covers of the space $[0, 1]$, while trying to avoid any that have a finite subcover. We could begin with $]\frac{1}{4}, \frac{3}{4}[$, then join $]\frac{1}{8}, \frac{7}{8}[$, followed by $]\frac{1}{16}, \frac{15}{16}[, \ldots$, and so on. The union of the infinite sequence of intervals

$$U_n =]2^{-n}, 1 - 2^{-n}[, \quad (n = 1, 2, 3, \ldots)$$

is the open interval $]0, 1[$. It is obvious that there is no way to reach the same union with only a finite number of these intervals. But we still need to cover 0 and 1. We can add two more intervals, both open sets in the space $[0, 1]$: for example $[0, 10^{-100}[$ and $]1 - 10^{-100}, 1]$. But now we may drop all the sets U_n, except just one, and still cover $[0, 1]$.

It is the same however we try to cover $[0, 1]$ with open sets; one may always drop all but a finite number of them and still reach the same union.

4.2.2 Compact Spaces and Compact Sets

Let (X, d) be a metric space and let $A \subset X$. We say that A is a compact set if the space (A, d) is compact. This is a standard way to transfer predicates applicable to metric spaces to sets in metric spaces.

If A is a compact set then every covering of A with sets *open relative to A* has a finite subcovering. But often it is irksome to look at sets open relative to A; we would much rather look at sets open in X. What if we cover A with sets open in X? These sets may go outside A. So we are saying that a family $(U_i)_{i \in I}$ of sets open in X satisfies $A \subset \bigcup_{i \in I} U_i$. Can we find a finite subfamily with the same property? In fact we have the following result that eases the study of compact sets.

Proposition 4.6 *Let (X, d) be a metric space and $A \subset X$. Then A is a compact set if and only if the following condition is satisfied: if $(U)_{i \in I}$ is a family of sets open in X and $A \subset \bigcup_{i \in I} U_i$ then there exists a finite set $\{i_1, \ldots, i_n\} \subset I$, such that $A \subset \bigcup_{k=1}^{n} U_{i_k}$.*

Proof Suppose that A satisfies the given condition. We show that A is a compact set. Let $A = \bigcup_{i \in I} V_i$ where $V_i \subset A$ and is open relative to A for each i. Then for each i there exists a set U_i, open in X, such that $V_i = U_i \cap A$. Hence $A \subset \bigcup_{i \in I} U_i$. There exists a finite subfamily $(U_{i_k})_{k=1}^{n}$, such that $A \subset \bigcup_{k=1}^{n} U_{i_k}$. But then we also have $A = \bigcup_{k=1}^{n} V_{i_k}$. We conclude that A is a compact set.

Conversely, suppose that $A \subset X$ is a compact set and let us show that it satisfies the condition stated in the proposition. Let $A \subset \bigcup_{i \in I} U_i$ where each set U_i is open in X. Now $A = \bigcup_{i \in I}(U_i \cap A)$ and each set $U_i \cap A$ is open relative to A. But then there exists a finite set $\{i_1, \ldots, i_n\}$ such that $A = \bigcup_{k=1}^{n}(U_{i_k} \cap A)$, which implies $A \subset \bigcup_{k=1}^{n} U_{i_k}$. $\qquad\qquad\square$

Proposition 4.7 *Let (X, d) be a metric space and $A \subset X$. Then*

1) If A is a compact set then A is closed in X and bounded.

2) If X is compact then A is compact if and only A is closed in X.

Proof
1) Let A be compact. First we show that A is closed. If $A = X$ then A is certainly closed, so that we assume that $A \neq X$. Let $p \in A^c$. For each $x \in A$ there exists $r_x > 0$, such that

$$B_{r_x}(x) \cap B_{r_x}(p) = \emptyset.$$

The balls $(B_{r_x}(x))_{x \in A}$ cover A; so there exists a finite set $\{x_1, \ldots, x_n\} \subset A$, such that the balls $B_{r_{x_k}}(x_k)$, $(k = 1, \ldots, n)$, cover A. Let

$$r = \min(r_{x_1}, \ldots, r_{x_n}).$$

The ball $B_r(p)$ does not intersect A, since it meets none of the balls $B_{r_{x_k}}(x_k)$, $(k = 1, \ldots, n)$, which cover A. We conclude that A^c is open so that A is closed.

Next we show that A is bounded. Let $a \in X$. We have

$$A \subset \bigcup_{n=1}^{\infty} B_n(a).$$

The family $(B_n(a))_{n=1}^{\infty}$ of open balls is increasing and covers A. Therefore, a finite subfamily covers A. Hence there exists n, such that $A \subset B_n(a)$, that is, A is bounded.

2) Suppose that X is compact. If A is compact then A is closed, as we have just seen. Conversely suppose that A is closed. Let $(U_i)_{i \in I}$ be a family of sets, open in X, such that

$$A \subset \bigcup_{i \in I} U_i.$$

The set A^c is open and the enlarged family

$$\{A^c\} \cup \{U_i : i \in I\}$$

covers X. A finite subfamily of this covers X, and therefore also A, but if this subfamily contains A^c we can obviously throw it out and A will still be covered. We conclude that a finite subfamily of $(U_i)_{i \in I}$ covers A, and so A is compact. $\square$

4.2.3 Continuous Functions on Compact Spaces

We come to one of the most important properties of compact spaces.

Proposition 4.8 *Let X be a compact metric space, let Y be a metric space, and let $f : X \to Y$ be continuous. Then $f(X)$ is a compact set in Y.*

Proof Let $(U_i)_{i \in I}$ be a family of sets, each open in Y, such that

$$f(X) \subset \bigcup_{i \in I} U_i.$$

Now $(f^{-1}(U_i))_{i \in I}$ is an open covering of X and therefore there exists a finite subfamily $(f^{-1}(U_{i_k}))_{k=1}^{n}$ that covers X. But then

$$f(X) \subset \bigcup_{k=1}^{n} U_{i_k}.$$

$\square$

The proposition has some spectacular corollaries, including the following very useful ones. The proof of the first should now be obvious. The second provides a case when the inverse of an injective continuous mapping is automatically continuous.

Proposition 4.9 *Let X be a compact metric space, let Y be a metric space, and let $f : X \to Y$ be continuous. Then $f(X)$ is bounded and closed. If $Y = \mathbb{R}$ then f attains its maximum and minimum.*

Proposition 4.10 *Let X be a compact metric space, let Y be a metric space, and let $f : X \to Y$ be continuous. Suppose that f is injective. Then the inverse mapping $f^{-1} : f(X) \to X$ is continuous.*

Proof Let $W \subset X$ be closed. Then W is compact, so that $f(W)$ is also compact, hence closed relative to $f(X)$. But $f(W) = (f^{-1})^{-1}(W)$, and the fact that this is closed relative to $f(X)$ whenever W is closed in X implies that f^{-1} is continuous on its domain $f(X)$ (by proposition 2.23). $\square$

4.2.4   Uniform Continuity

The topic of this section has been mentioned several times already in the context of real analysis.

Definition Let X and Y be metric spaces and let $f : X \to Y$. We say that f is uniformly continuous if the following condition is satisfied: for all $\varepsilon > 0$ there exists $\delta > 0$, such that $d(f(x), f(x')) < \varepsilon$ for all x and x' in X that satisfy $d(x, x') < \delta$.

All mappings that satisfy a Lipschitz condition are uniformly continuous. The following result was mentioned earlier in the case of a cube in $\mathbb{R}^n$. It is a striking case where the open covering definition of compactness can be used to advantage. Note the common trick (in the proof) of halving the radii of the balls before choosing a finite subcovering. We shall see more examples of this in the coming pages.

Proposition 4.11 *Let X and Y be metric spaces, let $f : X \to Y$ be continuous and assume that X is compact. Then f is uniformly continuous.*

Proof Let $\varepsilon > 0$. For each $a \in X$ there exists $r_a > 0$, such that

$$d(f(x), f(a)) < \tfrac{1}{2}\varepsilon$$

for all x that satisfy $d(x, a) < r_a$. The family of open balls

$$(B_{r_a/2}(a))_{a \in X}$$

covers X, so by compactness there exists a finite set of points $\{a_1, \dots, a_n\}$, such that the balls

$$B_{r_{a_k}/2}(a_k), \quad (k = 1, 2, \dots n),$$

cover X. Let

$$\delta = \tfrac{1}{2} \min(r_{a_1}, \dots, r_{a_n}).$$

Suppose that x and x' satisfy $d(x, x') < \delta$. There exists $k \in \{1, \dots, n\}$, such that $x' \in B_{r_{a_k}/2}(a_k)$, that is, $d(x', a_k) < r_{a_k}/2$. This implies

$$d(f(x'), f(a_k)) < \tfrac{1}{2}\varepsilon.$$

Since $d(x, x') < \delta < r_{a_k}/2$ we find $d(x, a_k) \le d(x, x') + d(x', a_k) < r_{a_k}$. Hence we also have

$$d(f(x), f(a_k)) < \tfrac{1}{2}\varepsilon.$$

Taking the inequalities together we find $d(f(x), f(x')) < \varepsilon$. $\square$

4.2.5   *Exercises*

1. Show that the union of a finite family of compact sets in a metric space is compact.
2. Show how using proposition 4.10 can simplify the proof of proposition 3.6 (which builds a homeomorphism from Cantor's middle thirds set B to the product space $2^{\mathbb{N}+}$).
3. Show that a metric space is compact if and only if it has the following property, called the *finite intersection property*: if $(F_i)_{i \in I}$ is a family of closed sets in X, such that $\bigcap_{i \in I} F_i = \emptyset$, then there exists a finite subfamily $(F_{i_k})_{k=1}^n$, such that $\bigcap_{k=1}^n F_{i_k} = \emptyset$.
4. Let $(A_n)_{n=1}^\infty$ be a sequence of compact sets in a metric space X, such that none of them is empty and $A_{n+1} \subset A_n$ for each n. Show that $\bigcap_{n=1}^\infty A_n \neq \emptyset$.

   Hint. Use the previous exercise.
5. Let X and Y be metric spaces and let X be compact. Let $f : X \to Y$ be continuous and let $A \subset X$. Suppose that $f(A^c) \subset (f(A))^c$ (that is, f maps the complement of A into the complement of $f(A)$).

   a) Show that f maps sets closed relative to A to sets closed relative to $f(A)$.
   b) Suppose in addition that the restriction $f|_A$ of f to A is injective. Show that $f|_A$ is a homeomorphism of A on to $f(A)$.

6. Let $f : \mathbb{R} \to \mathbb{C}$ be the function $f(t) = e^{2\pi i t}$. Show that the restriction of f to the interval $]0, 1[$ is a homeomorphism of $]0, 1[$ on to the unit circle minus one point, $T \setminus \{1\}$.

   Hint. Use the previous exercise.
7. Prove Dini's theorem: if M is a compact metric space, $(f_n)_{n=1}^\infty$ an increasing sequence of continuous real functions with domain M, converging pointwise to a continuous function g, then f_n converges to g uniformly.

   Hint. Use the reformulations of pointwise convergence and uniform convergence set out in 3.3 exercise 4.
8. Let X be a compact metric space and let $(U_i)_{i \in I}$ be an open covering of X. Show that there exists $\ell > 0$ such that, if A is a subset of X with diameter less that ℓ, there exists i such that $A \subset U_i$.

   Hint. We can assume the open cover is finite, $U_1, \ldots, U_n$. For each x choose $r(x) > 0$ and $k(x)$ in $[[1, n]]$, such that $B_{r(x)}(x) \subset U_{k(x)}$. Cover X by finitely many balls of the family $B_{r(x)/2}(x)$, $(x \in X)$. Figure out how to define ℓ.

   Note. A number ℓ with the property stated here is sometimes called a Lebesgue number for the covering $(U_i)_{i \in I}$.

4.3   Equivalence of Compactness and Sequential Compactness

The preceding row of propositions show that compact spaces and sequentially compact spaces share some highly significant properties. We begin to narrow the gap between the two notions by approaching the Bolzano-Weierstrass property from the compact side.

Proposition 4.12 *Let (X, d) be a compact space and let $A \subset X$ be an infinite subset. Then A has at least one limit point in X.*

Proof Assume, to the contrary, that no point of X is a limit point of A. For each $x \in X$ there exists an open set U_x, such that $x \in U_x$ but U_x contains no point of A, except possibly x itself (should x be in A). The family $(U_x)_{x \in X}$ is an open cover of X and hence it has a finite subfamily $(U_{x_k})_{k=1}^n$ that covers X. But then A must be finite, since the union of the subfamily contains at most n points that can be in A. $\square$

The property of a space, that every infinite set has a limit point, is sometimes called limit point compactness. It is yet another version of compactness to stand beside compactness itself and sequential compactness. The proposition says that compactness implies limit point compactness.

It is rather easy to see that sequential compactness and limit point compactness are equivalent. We summarise the proofs in two convoluted sentences that the reader is invited to unwrap. Given limit point compactness and a sequence $(x_n)_{n=1}^\infty$, we observe that either the sequence has only a finite number of distinct terms, in which case it has a constant, hence convergent, subsequence; or else the set of terms is infinite, has therefore a limit point a, and a subsequence with limit a is obtained by defining inductively an increasing sequence $(k_n)_{n=1}^\infty$, such that $x_{k_n} \in B_{1/n}(a)$. Conversely given sequential compactness and an infinite set A, we find a sequence $(x_n)_{n=1}^\infty$ in A that has distinct terms; then a limit a of a convergent subsequence of $(x_n)_{n=1}^\infty$ is a limit point of A.

In order to close the remaining gap between compactness and sequential compactness we need another notion.

Definition A metric space (X, d) is said to be *totally bounded* if, for all $\varepsilon > 0$, there exists a covering of X that consists of finitely many balls of radius ε.

The property of total boundedness can be described in the following pleasant and graphic fashion as a matter of town planning. Think of X as a town with the metric being geographical distance. For every $\varepsilon > 0$ one may place a finite number of bus stops, so that nobody has to walk further than ε to catch a bus.

Now we can state the equivalence theorem. It provides us in condition 3 with an extremely useful necessary and sufficient condition for compactness of a metric space, which one reaches for repeatedly. The remarkable thing is that neither of the two conditions comprising condition 3 is a topological property.

Proposition 4.13 *Let (X, d) be a metric space. The following are equivalent:*

1) X is compact.

2) X is sequentially compact.

3) X is complete and totally bounded.

Proof

$1 \Rightarrow 2$. The proof was described above.

$2 \Rightarrow 3$. Assume that X is sequentially compact. Let $(a_n)_{n=1}^{\infty}$ be a Cauchy sequence in X. There exists a convergent subsequence $(a_{k_n})_{n=1}^{\infty}$; call its limit b. We shall show that $a_n \to b$.

Let $\varepsilon > 0$. There exists N, such that $d(a_n, a_m) < \varepsilon$ for all $n \geq N$ and $m \geq N$. We can find $m \geq N$ (in the sequence k_i), such that $d(a_m, b) < \varepsilon$. But then for all $n \geq N$ we have $d(a_n, b) \leq d(a_n, a_m) + d(a_m, b) < 2\varepsilon$. We conclude that $a_n \to b$. Hence X is complete.

Next we show that X is totally bounded. We do this by contradiction. Suppose X is not totally bounded. There exists $\varepsilon > 0$, such that there exists no covering of X by finitely many balls of radius ε. Let $a_1 \in X$. By assumption the ball $B_\varepsilon(a_1)$ alone does not cover X. Hence there exists $a_2 \notin B_\varepsilon(a_1)$. Again the union of the two balls $B_\varepsilon(a_1) \cup B_\varepsilon(a_2)$ does not cover X. Hence there exists $a_3 \notin B_\varepsilon(a_1) \cup B_\varepsilon(a_2)$. Observe that $d(a_1, a_2)$, $d(a_1, a_3)$ and $d(a_2, a_3)$ are all greater than ε.

Continuing in the same way, we can define inductively a sequence $(a_n)_{n=1}^{\infty}$ with the property that $d(a_n, a_m) \geq \varepsilon$ for all n and m such that $n \neq m$. Here's how. Given that a_i has been defined for $i = 1, 2, \ldots, n$ with the property that $d(a_i, a_j) \geq \varepsilon$ for all $i, j \leq n$ such that $i \neq j$, we can produce a_{n+1} such that

$$a_{n+1} \notin \bigcup_{i=1}^{n} B_\varepsilon(a_i),$$

because the union does not cover X by assumption.

Now it is clear that the sequence $(a_n)_{n=1}^{\infty}$ cannot have a convergent subsequence. This contradicts the assumption that X is sequentially compact and completes the proof that condition 2 implies condition 3.

$3 \Rightarrow 1$. Assume that X is complete and totally bounded. Again we seek a proof by contradiction and assume that X is not compact. There exists an open covering $(U_i)_{i \in I}$ of X, such that no finite subfamily covers X.

Because X is totally bounded, there exists, for each integer $n = 1, 2, 3, \ldots$, a finite covering of X by balls of radius 2^{-n}. We refer to them as covering number 1, 2, 3, and so on. Consider the balls in covering number 1. For at least one of them, say $B_{1/2}(a_1)$, it is the case that it is not included in the union of any finite subfamily of $(U_i)_{i \in I}$. For if each of the balls in covering number 1 was included in the union of a finite subfamily of $(U_i)_{i \in I}$, we could piece together these subfamilies to produce a finite subfamily of $(U_i)_{i \in I}$ that covered all of X.

Consider next covering number 2. For at least one of them, say $B_{1/4}(a_2)$, it is the case that it is not included in the union of any finite subfamily of $(U_i)_{i \in I}$, *but in addition we can choose $B_{1/4}(a_2)$ so that it intersects $B_{1/2}(a_1)$*. For if we cannot thus choose $B_{1/4}(a_2)$, it would be possible to cover $B_{1/2}(a_1)$ contrary to its construction in the following way: for each ball in covering number 2 that meets $B_{1/2}(a_1)$ we choose a finite subfamily of $(U_i)_{i \in I}$ that covers it. On uniting these subfamilies we get a finite subfamily of $(U_i)_{i \in I}$ that covers $B_{1/2}(a_1)$; but such a finite subfamily does not exist by the previous paragraph.

Continuing the arguments of the last two paragraphs we build a sequence of balls $B_{2^{-n}}(a_n)$, $n = 1, 2, 3, \ldots$, such that none of them can be covered by a finite subfamily of $(U_i)_{i \in I}$, and such that each intersects its predecessor, that is,

$$B_{2^{-n}}(a_n) \cap B_{2^{-n-1}}(a_{n+1}) \neq \emptyset.$$

The precise inductive definition of this sequence is along these lines. When we have defined $B_{2^{-j}}(a_j)$ for $j = 1, 2, \ldots, n$, we look into the covering number $n + 1$, (by open balls of radius 2^{-n-1}), and choose one of them, $B_{2^{-n-1}}(a_{n+1})$, that cannot be covered by a finite subfamily of $(U_i)_{i \in I}$, and such that it intersects $B_{2^{-n}}(a_n)$.

The next step is to show that the centres of the balls, the sequence $(a_n)_{n=1}^{\infty}$, form a Cauchy sequence. Using the fact that each ball meets its predecessor, and applying the triangle inequality, we see that for $m < n$ we have:

$$d(a_m, a_n) \leq d(a_m, a_{m+1}) + d(a_{m+1}, a_{m+2}) + \cdots + d(a_{n-1}, a_n)$$
$$\leq (2^{-m} + 2^{-m-1}) + (2^{-m-1} + 2^{-m-2}) + \cdots + (2^{-n+1} + 2^{-n})$$
$$< 3 \cdot 2^{-m}$$

We conclude that $(a_n)_{n=1}^{\infty}$ is a Cauchy sequence, and hence has a limit b. But there exists $j \in I$, such that $b \in U_j$. There exists $\varepsilon > 0$, such that $B_\varepsilon(b) \subset U_j$. For sufficiently large n we have

$$B_{2^{-n}}(a_n) \subset B_\varepsilon(b) \subset U_j.$$

Actually it is enough to choose n so that $2^{-n} < \varepsilon/2$ and $d(a_n, b) < \varepsilon/2$. Then $B_{2^{-n}}(a_n)$ is a subset of U_j, when by construction it was not supposed to be possible to cover $B_{2^{-n}}(a_n)$ by finitely many sets from the family $(U_i)_{i \in I}$. And we have succeeded with just one! This contradiction concludes the proof. □

The proof just given that condition 3 implies condition 1 is a fairly complicated proof by contradiction, of the *reductio ad absurdum* type. It is useful to try to define the contradictory conclusion at the end of the deductive chain that commences with the assumption $\neg(1) \wedge (3)$. The deductive chain terminates with a statement p which asserts that two objects exist that together have contradictory properties. In this case the proposition p tells us that there exist a family of open sets and a ball, such that:

(a) the ball is not included in the union of any finite subfamily; and (b) the ball is included in one element of the family.

The conclusion p is logically untrue. It belongs to a general type of logically untrue statement that arises frequently in proofs by contradiction. The general form asserting the existence of n objects that together have contradictory properties is:

$$(\exists A_1, \ldots, A_n)\big(F(A_1, \ldots, A_n) \wedge \neg F(A_1, \ldots, A_n)\big)$$

where $F(A_1, \ldots, A_n)$ is a predicate taking n arguments of the right kind. Thus $F(A_1, A_2)$ could be 'the family A_1 has a finite subfamily that covers the ball A_2'.

Another interesting point about the proof is the role of the axiom of choice. In the proof that (2) implies (3), more precisely where it is shown that sequential compactness implies total boundedness, a sequence of points is built by taking one from each of a sequence of non-empty subsets of X, the latter sequence being defined inductively. In the proof that (3) implies (1), it is given that for each n there exists a covering of X by finitely many balls of radius 2^{-n}. One such covering is taken for each n from a potentially infinite number of possibilities. Again, from covering number n we are asked to select a ball that has a given property, which does not necessarily define that ball uniquely; and perform this for each n inductively.

All these examples call for the axiom of choice. The reader with a bent for set theory might supply the missing arguments in detail. For most practitioners of analysis, the proofs we provided are quite sufficient, as either they know that the missing details can be supplied using the axiom of choice, or else they are happy to remain in blissful ignorance of the fact. The interested reader might scan the other proofs in the section for tacit uses of the axiom of choice.

Using proposition 4.13 we obtain a key theorem of multivariate analysis:

Proposition 4.14 (Heine-Borel Theorem) *A set $A \subset \mathbb{R}^n$ is compact if and only if it is closed and bounded.*

Proof We have seen that being closed and bounded is necessary and sufficient for A to be sequentially compact, and we apply the previous proposition.   □

The reader may have seen a proof of the Heine-Borel theorem by bisection (perhaps more correctly called dissection). It is interesting to recall it because of the light it sheds on the complicated proof in proposition 4.13 that condition 3 implies condition 1.

It is assumed that A is a cube T in $\mathbb{R}^n$ (the general case of Heine-Borel follows easily from this one). Now T is dissected by planes parallel to the coordinate planes $x_i = 0$ into small cubes with side 2^{-k} for $k = 1, 2, 3, \ldots$ and so on. This produces a sequence of dissections into ever smaller subcubes. Again pursuing a contradiction we suppose that an open covering $(U_i)_{i \in I}$ is known that has no finite subcovering. Then for each k we can find a subcube of side 2^{-k} that is not included in the union of a finite subfamily of $(U_i)_{i \in I}$. What is more we can choose this subcube so that it is a subset of the previous subcube, the one with side 2^{-k+1}. These subcubes converge to a point in T, and the fact that this point lies in one of the sets U_i leads to a contradiction.

We recall this argument because of a strong parallel with the proof that condition 3 implies condition 1 in proposition 4.13. In that case we did not have at our disposal the geometry that can be used in $\mathbb{R}^n$ to dissect sets into smaller parts. Instead we had to use the paraphernalia of total boundedness, and the corresponding finite coverings by balls of radius 2^{-k}. The way these were used is quite analogous to the proof of Heine-Borel in $\mathbb{R}^n$ by dissection.

The original Heine-Borel theorem did not mention compactness of course. It stated more or less that an open covering of a cube had a finite subcovering. The earliest intimations of it assumed a one-dimensional setting, the cube was an interval, and the covering sets also intervals. The full theorem was the fruit of efforts of many mathematicians working in the last decades of the 19th century.

4.3.1 Relative Compactness

A set A in a metric space X is said to be relatively compact when its closure $\overline{A}$ is compact. The concept is useful because often we would like to know that every sequence in A has a convergent subsequence but we don't mind if the limit is outside A. The term 'relatively' used here is a little troublesome; "relative to what?" one would like to ask. It is therefore not surprising that other terms have been used, for example 'precompact', about the use of which there is some confusion—it is sometimes a synonym for 'totally bounded'.

Relative compactness is a property of sets in a metric space. Thus the set $]0, 1[$ is relatively compact in $\mathbb{R}$, but not relatively compact in $]0, 1[$.

Proposition 4.15 *Let X be a metric space and let $A \subset X$. Then the following are equivalent:*

1) *A is relatively compact.*
2) *Every sequence in A has a subsequence that converges in X.*
3) *A is totally bounded and $\overline{A}$ complete.*

Proof By proposition 4.13, we have that (1) implies (3). By the same, (3) implies (1), provided we can show that if A is totally bounded then so is $\overline{A}$. This is left to the reader as an exercise. The same proposition shows that (1) implies (2). It remains to show that (2) implies (1).

Suppose that every sequence in A has a subsequence that converges in X. We shall show that $\overline{A}$ is sequentially compact. A sequence in $\overline{A}$ could lie wholly in A, it could have infinitely many terms in A, or it could have finitely many terms in A. The outstanding case is the last one, and we may as well assume that we have a sequence $(x_n)_{n=1}^{\infty}$ that lies wholly in $\overline{A} \setminus A$. For each n there exists $a_n \in A$, such that $d(a_n, x_n) < 1/n$. There exists a subsequence $(a_{k_n})_{n=1}^{\infty}$ that is convergent; its limit is in $\overline{A}$. But then $(x_{k_n})_{n=1}^{\infty}$ is also convergent and has the same limit. □

4.3.2   Local Compactness

We introduce a new topological property.

Definition A metric space is said to be locally compact if every point has a compact neighbourhood.

An equivalent formulation of local compactness that references the metric is: for all $x \in X$ there exists $r > 0$, such that the closed ball $B_r^-(x)$ is compact.

Exercise Prove the claim of the previous paragraph.

Probably the most important examples of locally compact spaces are the coordinate spaces $\mathbb{R}^n$, with any of the equivalent metrics $\|x - y\|_p$. The closed balls $B_r^-(a)$ are compact by the Heine-Borel theorem. In the usual way we extend the notion to sets in a metric space. A set A in a metric space X is locally compact when the subspace A is locally compact.

The supply of locally compact spaces is increased by the following result:

Proposition 4.16 *Let X be a locally compact space. Then all open subsets of X and all closed subsets of X are locally compact.*

Proof Let A be open in X and let $a \in A$. There exists $r > 0$, such that the closed ball $B_r^-(a)$ is compact, and $s > 0$, such that $B_s^-(a) \subset A$. Setting $t = \min(r, s)$ we see that $B_t^-(a)$ is a compact neighbourhood of a in A.

Let A be closed in X and let $a \in A$. There exists $r > 0$, such that the closed ball $B_r^-(a)$ is compact, and the intersection $B_r^-(a) \cap A$ is compact (since A is closed) and is a neighbourhood of a in A.   □

The open r-neighbourhood of a set A in a metric space X was defined in 2.2 exercise 24 to be the set

$$N_r(A) = \{x \in X : d(x, A) < r\}.$$

Proposition 4.17 *Let X be locally compact, let $M \subset X$ and suppose M is compact. Then there exists $r > 0$, such that the closure of the open r-neighbourhood of M, the set $\overline{N_r(M)}$, is compact.*

Proof For each $x \in M$ there exists $r_x > 0$, such that the closed ball $B_{r_x}^-(x)$ is compact. Since M is compact there exists a finite set of points $\{a_1, \ldots, .a_m\}$ in M such that

$$M \subset \bigcup_{k=1}^{m} B_{r_{a_k}/2}(a_k).$$

Let

$$W = \bigcup_{k=1}^{m} B_{r_{a_k}}^{-} (a_k).$$

Note that W is compact, being a finite union of compact sets.

Choose r so that $0 < r < \frac{1}{2} \min_{1 \le k \le m} (r_{a_k})$. Suppose that x satisfies $d(x, M) < r$. Then there exists $y \in M$, such that $d(x, y) < r$, and k, such that $d(y, a_k) < r_{a_k}/2$. By the triangle inequality we find

$$d(x, a_k) \le d(x, y) + d(y, a_k) < r + \frac{r_{a_k}}{2} < \frac{r_{a_k}}{2} + \frac{r_{a_k}}{2} < r_{a_k}.$$

Hence $x \in W$.

We deduce that $N_r(M) \subset W$, and since W is compact, we conclude that the closure of $N_r(M)$ is compact. □

4.3.3 Exercises

1. Let (X, d) be a metric space and let $A \subset X$. Show that the subspace (A, d) is totally bounded if and only if A satisfies the following condition: for each $\varepsilon > 0$ there exists a finite family $(x_j)_{j=1}^{n}$ in X, such that $A \subset \bigcup_{j=1}^{n} B_\varepsilon(x_j)$.

   Hint. To prove that the condition is sufficient move the bus stops inside the town boundary.

2. Let (X, d) be a metric space and $A \subset X$. Suppose that A is totally bounded. Show that $\overline{A}$ is totally bounded.

3. Show that the metric space (X, d) is totally bounded if and only if the following condition is satisfied: for each $\varepsilon > 0$ there exists a finite family of sets $(A_j)_{j=1}^{n}$, such that diam $A_j < \varepsilon$ for each j, and $X = \bigcup_{j=1}^{n} A_j$.

4. Let $A \subset \mathbb{R}^n$. Show that A is totally bounded if and only if it is bounded. Use this to give another proof of The Heine-Borel theorem by directly applying proposition 4.13 item 3.

   Hint. Do it first in the case that A is a cube and the metric is $\|x - y\|_\infty$.

5. Find a bounded subset of $A \subset \ell^2$ that is not totally bounded.

6. Show that a Cauchy sequence is totally bounded (that is, its set of terms is totally bounded).

7. We mentioned the property of limit point compactness, that every infinite set has a limit point, and pointed out that it is equivalent to sequential compactness (and therefore also compactness). Show that for a space to be limit point compact, it is sufficient that every countable infinite set has a limit point.

8. a) The Hilbert cube K (mentioned in section 4.1) is the set of all sequences $a = (a_n)_{n=1}^{\infty} \in \ell^2$, such that $0 \le a_n \le \frac{1}{n}$ for each n. Show that K is compact.

Hint. The tricky bit is to show that K is totally bounded. For this purpose you might like to define the set

$$K_n = \{x \in K : x_k = 0 \text{ for all } k > n\},$$

and identify K_n with a subset of $\mathbb{R}^n$.

b) Another set also commonly called the Hilbert cube is the set of all sequences $x = (x_n)_{n=1}^{\infty} \in \mathbb{R}^{\mathbb{N}+}$, such that $0 \le x_n \le 1$ for all n. It is denoted by H^{∞}. The space $\mathbb{R}^{\mathbb{N}+}$ has the metric

$$d(a, b) = \sum_{n=1}^{\infty} 2^{-n} \frac{|a_n - b_n|}{1 + |a_n - b_n|}$$

or any equivalent metric. Note that H^{∞} is the product $[0, 1]^{\mathbb{N}+}$, so it is aptly named.

Define the map $f : K \to H^{\infty}$, so that $f(x) = y$ means that $y_n = nx_n$ for each n. Show that f is bijective and continuous. Deduce that H^{∞} is compact and f is a homeomorphism.

9. Let X and Y be compact spaces. Prove that $X \times Y$ is compact by showing that every sequence in it has a convergent subsequence (along the lines of the proof of proposition 4.2).

10. Obviously the conclusion of the previous exercise extends to any finite cartesian product of compact spaces. Now extend it to a countable infinite product. Let $(X_n)_{n=1}^{\infty}$ be a sequence of compact metric spaces. Show that every sequence in the product space $\Pi_{n=1}^{\infty} X_n$ has a convergent subsequence. Conclude that the product space is compact.

Hint. Some modification of the proof of proposition 4.2 is required to ensure that after restricting to a subsequence infinitely many times there is anything left at all.

11. Let X be a metric space, let $p \in X$ and suppose that p has a *compact* neighbourhood W. Let $f : W \to [0, \infty[$ be a continuous function, such that p is the unique $x \in W$ such that $f(x) = 0$. Show that the sets

$$A_r := \{x \in X : f(x) < r\}, \quad (r > 0)$$

form a *base for the neighbourhoods of* p. The meaning attached to this notion is as follows: a family $(A_i)_{i \in I}$ of sets is a base for the neighbourhoods of p if every A_i is a neighbourhood of p, and for all neighbourhoods U of p there exists i, such that $A_i \subset U$.

12. Some knowledge of algebra is assumed in this exercise, which states some examples, useful in multivariate analysis, of the result of the previous one.

a) Let $Q(x)$ be a positive definite quadratic form in n real variables. Show that the sets

$$A_r := \{x \in \mathbb{R}^n : Q(x) < r\}, \quad (r > 0)$$

form a base for the neighbourhoods of 0 in $\mathbb{R}^n$.
b) Let $F : \mathbb{R}^n \to \mathbb{R}$ be continuous, let $a \in \mathbb{R}^n$ and let F have a local strict minimum at a. Then for suitably chosen $s > 0$ and $t > 0$, the sets

$$A_r := \{x \in \mathbb{R}^n : F(x) < F(a) + r\} \cap B_t(a), \quad (0 < r < s)$$

form a base for the neighbourhoods of the point a.

13. Let X be a compact space and let Δ be the *diagonal*:

$$\Delta = \{(x, y) \in X \times X : x = y\}.$$

Show that for every set U open in $X \times X$, such that $\Delta \subset U$, there exists $\delta > 0$, such that $(x, y) \in U$ whenever $d(x, y) < \delta$.

14. Let D be a countable set and let $(f_n)_{n=1}^\infty$ be a sequence of functions from D to a compact metric space Y. Show that the sequence $(f_n)_{n=1}^\infty$ has a subsequence that is pointwise convergent.
Hint. The space of functions from D to Y is the product space Y^D and you can apply the result of exercise 10.

15. Let X be a compact metric space and let $(W_i)_{i \in I}$ be a family of closed sets whose intersection M is non-empty.

a) Show that for each open set U such that $M \subset U$ there exists a finite subfamily $(W_{i_k})_{k=1}^n$ such that

$$\bigcap_{k=1}^n W_{i_k} \subset U$$

In particular, for each $\varepsilon > 0$ there exists a finite subfamily $(W_{i_k})_{k=1}^n$ such that

$$\bigcap_{k=1}^n W_{i_k} \subset N_\varepsilon(M)$$

where $N_\varepsilon(M)$ is the open ε-neighbourhood of M (see 2.2 exercise 24 and the present section).
b) Suppose that the family is a decreasing sequence $(W_i)_{i \in \mathbb{N}_+}$ (that is, $W_{i+1} \subset W_i$ for each i). Show that for each $\varepsilon > 0$ there exists i such that $W_i \subset N_\varepsilon(M)$.

16. Let M and U be subsets of a metric space X, such that M is compact, U is open and $M \subset U$. Show that there exists $r > 0$ such that

$$M \subset N_r(M) \subset U.$$

Hint By 4.1 exercise 4, $\inf\{d(x, y) : x \in M, \ y \in U^c\}$ is positive.

17. Let X be locally compact. Let M and U be subsets of X such that M is compact, U is open and $M \subset U$.

a) Show that there exist a compact set N and an open set V, such that

$$M \subset V \subset N \subset U.$$

b) ($\Diamond$) Show that there exists a continuous mapping $f : X \to [0, 1]$ such that $f = 1$ in M, and the support (2.3 exercise 16) of f is a compact subset of U.

Hint. See section 3.6.

18. ($\Diamond$) The results of the previous exercise allow one to obtain an important type of *partition of unity* (see 3.6 exercises 4 and 5) that is based on functions with compact support.

Let X be locally compact, let M be a compact set in X and let $(U_k)_{k=1}^n$ be a sequence of open sets such that

$$M \subset \bigcup_{k=1}^{n} U_k.$$

Prove that there exist continuous functions $f_k : X \to [0, 1], (k = 1, \dots, n)$, with compact support, such that

$$\operatorname{supp}(f_k) \subset U_k, \quad (k = 1, \dots, n)$$

$$\sum_{k=1}^{n} f_k(x) \leq 1, \quad (x \in X)$$

and

$$\sum_{k=1}^{n} f_k(x) = 1, \quad (x \in M).$$

19. Show that the Hausdorff pseudometric (section 1.2) on the set of all non-empty subsets of a metric space X becomes a metric when restricted to the set of all compact sets in X, and what is more, takes only finite values.

20. Let X be a metric space and assume that it is not compact. Show that X has a countably infinite discrete subspace A, such that A is closed in X.

21. a) Let X be a set. Show that all functions $f : X \to \mathbb{R}$ are bounded if and only if X is finite.

   b) ($\Diamond$) Show that a metric space X is compact if and only if all continuous functions $f : X \to \mathbb{R}$ are bounded.

   Hint. Use exercise 20 and Tietze's extension theorem, proposition 3.18.

22. Let E be a complete normed space. Show that a bounded set $A \subset E$ is relatively compact if and only if it satisfies the following condition: for all $\varepsilon > 0$ there exists a finite-dimensional vector subspace F of E, such that $d(x, F) < \varepsilon$ for all $x \in A$; or, equivalently, $A \subset N_\varepsilon(F)$. Check that this is satisfied for the Hilbert cube.

23. Let $1 \le p < \infty$. Show that a subset K of ℓ^p is compact if and only if it is closed and bounded, and satisfies in addition the following uniformity condition: for all $\varepsilon > 0$ there exists N, such that for all $a = (a_j)_{j=1}^\infty \in K$ we have $\sum_{j=N}^\infty |a_j|^p < \varepsilon$. Compare the Hilbert cube.

24. Let X be a complete normed space and let $A \subset X$. Show that if A is relatively compact then its convex hull (2.2 exercise 20) is relatively compact.

An interesting and useful application of the Heine-Borel theorem for the interval $[a, b]$ can be made to the study of so-called regulated functions. Let E be a normed space. A function $f : [a, b] \to E$ is said to be regulated if it has left and right-hand limits (not necessarily equal) at each point of the open interval $]a, b[$ as well as a left limit at b and right limit at a. For more on regulated functions see section 5.1.1.

We say that $f : [a, b] \to E$ is a step function if there exists a partition $(t_0, t_1, \ldots, t_m)$ of $[a, b]$ and vectors $c_0, \ldots, c_{m-1}$ in E, such that $f(s) = c_k$ for $s \in]t_k, t_{k+1}[$, $(k = 0, 1, \ldots m - 1)$. In other words, f is constant on each open subinterval of the partition.

All that was said in the previous two paragraphs can be stated with a general metric space replacing the normed space E. This is also true of the following exercise, although we cannot drop completeness for the sufficiency. The reader is referred to 3.3 for material concerning uniform convergence, including 3.3 exercise 12.

25. Let E be a complete normed space. Show that a function $f : [a, b] \to E$ is regulated if and only if it is the uniform limit of a sequence of step functions.

   Hint. (*Necessity*) Let f be a regulated function and let $\varepsilon > 0$. Using proposition 2.28 (adapted to one-sided limits), and the Heine-Borel theorem, show that there is a partition $(t_0, t_1, \ldots, t_m)$ of $[a, b]$, such that the oscillation of f in each open subinterval $]t_k, t_{k+1}[$ is less than ε. Then build a step function that approximates f uniformly with error less than ε. (*Sufficiency*) Again use proposition 2.28. Consult the proof of proposition 3.8.

The following exercises introduce spaces of complex analytic functions. Some knowledge of complex analysis, in particular of Cauchy's integral formula, is assumed.

Let D be the unit disc in the complex plane $\mathbb{C}$, that is,

$$D = \{z \in \mathbb{C} : |z| < 1\}.$$

The *disc algebra* $A(D)$ is the algebra of all complex analytic functions $f : D \to \mathbb{C}$, such that f extends to a continuous function on the closed disc $\overline{D}$. It is a complex vector space, and an algebra, where multiplication of two functions is defined pointwise. It is a normed space, bearing the norm

$$\|f\|_\infty = \sup_{\overline{D}} |f|.$$

In this manner $A(D)$ is a subspace of (complex) $C(\overline{D})$.

26. Now show that $A(D)$ is a Banach space, that is, it is complete.

   Hint. You will need to know that the uniform limit of a sequence of functions, all of which are analytic in a disc, is itself analytic. This is a consequence of Morera's theorem of complex analysis.

The space of functions analytic in an open set can be given a metric, similar to the treatment of the space $C^\infty(\mathbb{R})$ (see section 3.3). The analyticity of the functions makes for a quite different experience.

Let A be an open set in the complex plane $\mathbb{C}$. We denote the space of all complex analytic functions in A by $O(A)$. The metric assigned to $O(A)$ ascribes the following meaning to the notion that a sequence $(f_n)_{n=1}^\infty$ of analytic functions in A converges to a function g:

For all compact $K \subset A$ we have $f_n \to g$ uniformly in K.

To describe a suitable metric we let $(V_n)_{n=1}^\infty$ be an increasing sequence of bounded open sets in $\mathbb{C}$, such that $\bigcup_{n=1}^\infty \overline{V_n} = A$. Sets with these properties can be built using 2.3 exercise 12.

In particular cases we can describe V_n explicitly. Examples are $A = \mathbb{C}$, where we can take V_n to be the open disc of radius n and centre 0; and $A = D$ (the open unit disc), where V_n is the open disc of radius $1 - \frac{1}{n}$.

We may use as a metric

$$d(f, g) = \sum_{n=1}^\infty 2^{-n} \min\left(\sup_{\overline{V_n}} |f - g|, 1\right).$$

27. Show that a sequence $(f_n)_{n=1}^\infty$ in $O(A)$ converges to a function g with respect to the given metric if and only if $f_n \to g$ uniformly on every compact $K \subset A$.

   Note. This shows that the topology (section 2.6) induced on $O(A)$ is independent of the way the sets $(V_n)_{n=1}^\infty$ are chosen.

28. Show that $O(A)$ is complete with the given metric.

   Hint. Same as for exercise 26.

   Topological spaces were defined in 2.6. We can consider *compact topological spaces*, that are not necessarily metric spaces. The definition of compactness is clear: a topological space is said to be compact when every open covering has a finite subcovering. However, it is no longer true that compactness and sequential compactness are equivalent. First of all, we shall see an example of a sequentially compact space that is not compact:

29. a) Let X be a well-ordered set with a maximum element. Show that X is compact with its order topology (see 2.6 for the definition of order topology) and also sequentially compact.

   Hint. Let ω be the maximum element. For compactness let $(U_i)_{i \in I}$ be a covering of X by open sets, and consider the minimum y such that the interval $[y, \omega]$ is covered by finitely many sets of the covering. For sequential compactness apply the result of 2.6 exercise 5(b).

   b) Let X be a well-ordered set without a maximum element. Show that X is not compact with its order topology.

   Hint. Cover X with the sets $]-\infty, y[$, $(y \in X)$.

   We shall accept that there exists an uncountable well-ordered set W, such that every initial segment is countable. This was mentioned in 2.6 exercise 13.

   c) Show that W is sequentially compact with its order topology, but not compact.

An example of a compact space that is not sequentially compact is the space

$$X := \{0, 1\}^{2^{\mathbb{N}+}}$$

or the product of $2^{\mathbb{N}+}$ copies of the two-point space $2 := \{0, 1\}$. More concretely, since $2^{\mathbb{N}+}$ is the set of all sequences of the digits 0 and 1, we can view X as the *space of all functions that assign to each sequence of the digits 0 and 1 an element of the compact space* $\{0, 1\}$. The product topology was defined in 3.2 exercise 6. It is enough to know that a sequence of functions from $2^{\mathbb{N}+}$ to $\{0, 1\}$ is convergent when it converges pointwise. That X is compact results from Tychonov's theorem, that asserts that a product of compact spaces is compact with its product topology. There is no space to prove that in this text. It is easy to produce a sequence in X with no convergent subsequence:

30. Let $f_n : 2^{\mathbb{N}+} \to \{0, 1\}$ be such that $f_n(x) = x_n$; that is, f_n assigns to each sequence x its nth term. We view f_n as an element of X. Given a strictly increasing sequence $(k_n)_{n=1}^{\infty}$ of integers, produce an element y of $2^{\mathbb{N}+}$, such that

$$f_{k_n}(y) = \tfrac{1}{2}(1 + (-1)^n).$$

   Deduce that the sequence $(f_{k_n})_{n=1}^{\infty}$ does not converge in X.

4.4   Finite Dimensional Normed Vector Spaces

Compactness of the closed unit ball in $\mathbb{R}^n$, essentially the Heine-Borel theorem, is the key to establishing a decisive distinction between finite-dimensional normed spaces (in effect $\mathbb{R}^n$; the arena in which classical multivariate calculus is played out), and infinite-dimensional normed spaces, the most important of which are the function spaces, for example the so-called Sobolev spaces, a natural setting in which to study partial differential equations.

The first objective is to study finite-dimensional normed spaces. It turns out that, as far as the linear and the topological structures are concerned, a finite-dimensional normed space is determined by its dimension alone. If the dimension is n it is linearly homeomorphic to $\mathbb{R}^n$; the norm is immaterial.

We will write all our results for spaces over $\mathbb{R}$. The same conclusions hold for complex spaces, if we replace $\mathbb{R}^n$ by $\mathbb{C}^n$.

Proposition 4.18 *Let X be a finite-dimensional normed space with dimension n and equip $\mathbb{R}^n$ with the norm $\|a\|_1$. Then there exists a linear homeomorphism T : $\mathbb{R}^n \to X$.*

Proof Let $(u_1, \ldots, u_n)$ be a basis for X. Define $T : \mathbb{R}^n \to X$ by

$$Ta = a_1 u_1 + \cdots + a_n u_n$$

where $a = (a_1, \ldots, a_n)$. Then T is linear, and hence continuous by proposition 2.33.

Next we show that T^{-1} is continuous. By the Heine-Borel theorem the sphere $S = \{a : \|a\|_1 = 1\}$ in $\mathbb{R}^n$ is compact. Moreover for all $a \in S$ we have $Ta \neq 0$. Since T is continuous the image $T(S)$ is compact, hence closed, and moreover $0 \notin T(S)$. Hence there exists $\delta > 0$, such that $\|Ta\| \geq \delta$ for all $a \in S$.

Suppose now that $a \neq 0$. We have

$$\left\| T\left(\frac{a}{\|a\|_1} \right) \right\| \geq \delta,$$

so that

$$\|Ta\| \geq \delta \|a\|_1.$$

Let $a = T^{-1}v$ and deduce that $\|T^{-1}v\|_1 \leq \delta^{-1}\|v\|$ for all $v \in X$. That is, T^{-1} is continuous.   □

Proposition 4.19 *Let X and Y be normed spaces and suppose that X is finite-dimensional. Then all linear mappings $T : X \to Y$ are continuous. In particular two normed vector spaces of the same finite dimension are linearly homeomorphic.*

Proof Let X have dimension n and let $T_0 : \mathbb{R}^n \to X$ be a linear homeomorphism (as provided by proposition 4.18). Then $T = (T \circ T_0) \circ T_0^{-1}$ and this mapping is continuous since $T \circ T_0 : \mathbb{R}^n \to Y$ is continuous (by proposition 2.33) and T_0^{-1} is continuous.   □

Using this result we can prove a sweeping generalisation of a conclusion reached earlier, that all the norms $\|a\|_p$ on $\mathbb{R}^n$ are equivalent to each other.

Proposition 4.20 *Let X be a finite dimensional vector space over $\mathbb{R}$. Then all norms on X are equivalent and define the same open sets.*

Proof Consider two norms $\|x\|$ and $\|x\|'$. The identity mapping from X with the norm $\|\cdot\|$ to X with the norm $\|\cdot\|'$ is continuous, along with its inverse. Hence there exist constants $K > 0$ and $K' > 0$, such that

$$K\|x\| \leq \|x\|' \leq K'\|x\|$$

for all $x \in X$. The metrics $\|x - y\|$ and $\|x - y\|'$ are therefore equivalent, and define the same open sets.   □

Proposition 4.21 *All finite-dimensional normed spaces are complete.*

Proof If X is a finite-dimensional normed space there exists a linear homeomorphism $T : X \to \mathbb{R}^n$. It maps Cauchy-sequences in X to Cauchy-sequences in $\mathbb{R}^n$; but all Cauchy-sequences in $\mathbb{R}^n$ are convergent.   □

Proposition 4.22 *The Heine-Borel property, that the compact sets are precisely those sets that are bounded and closed, holds in all finite-dimensional normed vector spaces.*

The proof should be obvious. It is a remarkable fact, far from obvious, that in a normed space X the Heine-Borel property is both necessary and sufficient for the finite dimensionality of X. As it stands, the Heine-Borel property is not topological (it mentions bounded sets). However, we have already defined a topological property that, for normed spaces, is equivalent to the Heine-Borel property. We are alluding to *local compactness*.

A finite-dimensional normed space is locally compact because the closed ball $B_1^-(a)$ is a compact neighbourhood of a. We are going to prove a remarkable fact, that a normed vector space is finite-dimensional (an algebraic property) if and only if it is locally compact (a topological property). Hence in an infinite-dimensional normed space a closed and bounded set is not necessarily compact; some other condition has to be brought into play. What this other condition is in particular spaces is an important topic which goes mostly beyond the confines of the present text (but we shall touch upon it in section 4.5).

Nevertheless, we have already seen one example of a compact subset of an infinite-dimensional normed space, namely the Hilbert cube, in its manifestation as the set of all elements $a = (a_n)_{n=1}^{\infty}$ of ℓ^2 that satisfy $0 \leq a_n \leq \frac{1}{n}$ for each n (see 4.3 exercise 8). Two other examples will appear in this text: compact subsets of ℓ^p with $1 \leq p < \infty$ (see 4.3 exercise 23), and compact subsets of $C(M)$ for compact M (see 4.5).

Proposition 4.23 (Riesz's Lemma) *Let X be a normed space and let E be a proper, closed, linear subspace of X. Then for each α in the interval $0 < \alpha < 1$ there exists a vector v in X such that $\|v\| = 1$ and $d(v, E) > \alpha$.*

Proof Let $x \in X \setminus E$. Then $d(x, E) > 0$ since E is closed. Let $y \in E$ and set

$$v = \frac{x - y}{\|x - y\|}.$$

Let $0 < \alpha < 1$. We shall show that we may choose y so that v has the required properties. Note that $x - y \neq 0$ since $y \in E$ and $x \notin E$. For each $w \in E$ we have

$$y + \|x - y\| w \in E$$

and therefore

$$\|v - w\| = \left\| \frac{x - y}{\|x - y\|} - w \right\| = \frac{1}{\|x - y\|} \left\| x - y - \|x - y\| w \right\| \geq \frac{d(x, E)}{\|x - y\|}.$$

But then

$$d(v, E) \geq \frac{d(x, E)}{\|x - y\|}.$$

We only need to choose $y \in E$ to satisfy

$$\|x - y\| < \alpha^{-1} d(x, E).$$

This is possible because $\alpha^{-1} > 1$.   □

Proposition 4.24 *Let X be a normed space. Then the closed ball $B_1^-(0)$ is compact if and only if X is finite-dimensional.*

Proof We know that $B_1^-(0)$ is compact if X is finite-dimensional. That is Heine-Borel.

Now for the converse: suppose that $B_1^-(0)$ is compact. Then it is totally bounded and therefore there exist finitely many vectors $u_1, \ldots, u_n$, such that $B_1^-(0)$ is a subset of the union of the balls $B_{1/2}(u_k)$, $(k = 1, \ldots, n)$.

Let E be the subspace of X spanned by the vectors $u_1, \ldots, u_n$. Then E is finite-dimensional, hence complete (by proposition 4.21), hence closed in X. We shall show that $E = X$.

Suppose, if possible, that $E \neq X$. By Riesz's lemma there exists $v \in X$, such that $\|v\| = 1$ and $d(v, E) > \frac{1}{2}$. Then $v \in B_1^-(0)$, but v is not in any of the balls $B_{1/2}(u_k)$, $k = 1, \ldots, n$, because $\|v - u_k\| > \frac{1}{2}$ for each k. This is a contradiction because the balls were supposed to cover $B_1^-(0)$. We conclude therefore that $E = X$.   □

4.4.1 Exercises

1. a) Show that, for a normed vector space X, local compactness is equivalent to compactness of the closed unit ball $B_1^-(0)$.
   b) Show that a normed vector space is locally compact if and only if it is finite-dimensional.
2. Town planning on the surface of the unit ball $B_1(0)$ in an infinite-dimensional normed space X throws up a pleasant surprise. Infinitely many people could live on plots of identical size, in an area of finite diameter, without encroaching on their neighbours.

   Show that for each α in the interval $0 < \alpha < 1$ there exists an infinite sequence of unit vectors $(u_n)_{n=1}^\infty$, such that $\|u_i - u_j\| > \alpha$ whenever $i \neq j$.
3. Let X be a normed space and let E be a closed linear subspace of X. Does there exist a vector u with $\|u\| = 1$ and $d(u, E) = 1$? Riesz's lemma only tells us that $\sup_{\|u\|=1} d(u, E) = 1$, but is the supremum attained? The failure of Heine-Borel in infinite-dimensional spaces raises the possibility that it is not, but the question is not so straightforward. We explore this question in several cases. We begin with a fairly obvious point:

   a) Show that the supremum is attained for u, with $\|u\| = 1$, if and only if $\|u - x\| \geq 1$ for all $x \in E$.
   b) Let $X = \mathbb{R}^n$, E any proper linear subspace. Show that the supremum is attained. Note that any norm can be used here.
   c) Let $X = \ell^2$, $E = \{(a_n)_{n=1}^\infty \in \ell^2 : a_1 = 0\}$. Show that the supremum is attained.

   Define the following spaces, all Banach spaces, and each equipped with the norm $\|\cdot\|_\infty$:

   $$X = C[0, 1],$$
   $$Y = \{f \in X : f(0) = 0\},$$
   $$E = \{f \in X : \int_0^1 f = 0\},$$
   $$F = \{f \in Y : \int_0^1 f = 0\}.$$

   Explore whether the supremum is attained in each of the following cases:

   d) Y as a subspace of X.
   e) E as a subspace of X.
   f) F as a subspace of Y.
      Hint. If $\phi \in Y$ and $\|\phi\| = 1$ then $\int_0^1 |\phi| < 1$.

4. Let X be a normed space, let f be a continuous linear functional on X and let E be the kernel of f, that is, $E = f^{-1}(0)$.

   a) Show that

   $$|f(x)| = \|f\| \, d(x, E), \quad (x \in X).$$

   b) Show that the supremum in Riesz's lemma is attained for E if and only if there exists $x \in X$, such that $\|x\| = 1$ and $|f(x)| = \|f\|$.
   c) Compare 2.4 exercise 13 and produce another example where the supremum in Riesz's lemma is not attained, additional to the one(s) in the previous exercise.

4.5 ($\Diamond$) Ascoli

We have seen that bounded, closed sets in an infinite-dimensional normed space are not necessarily compact. In fact the closed unit ball is never compact in such a space. Some other property, in addition to boundedness and closedness, must come into play to guarantee compactness.

We have seen the example of the Hilbert cube (4.3 exercise 8), realised as the subspace K of ℓ^2 consisting of all real sequences $x = (x_k)_{k=1}^{\infty}$, that satisfy $0 \le x_k \le 1/k$ for all k. The proof that this is compact depends on the observation that, given $\varepsilon > 0$, there exists N, such that for all $x \in K$ we have $\sum_{n=N+1}^{\infty} x_k^2 < \varepsilon$. The error in approximating $\sum_{k=1}^{\infty} x_k^2$ by $\sum_{k=1}^{N} x_k^2$ can be made as small as we please by choosing a value of N common to all x in K. We can describe this in the following way: the convergence of the series $\sum_{k=1}^{\infty} x_k^2$ is uniform with respect to the sequence x in K. Other examples of a property, additional to boundedness, that guarantee total boundedness, were given in 4.3 exercises 22 and 23.

These examples provide a clue as to what we might seek in other cases. That the series $\sum_{k=1}^{\infty} x_k^2$ converges is the defining property of the space ℓ^2. Let M be a compact metric space and let us turn to the question of finding necessary and sufficient conditions for a subset $A \subset C(M)$ to be compact, The defining property, that distinguishes the functions that find themselves in $C(M)$ from those that don't, is continuity. If $A \subset C(M)$, we want a notion that ascribes to A uniformity of the continuity of elements of A. This has a name.

Definition Let M be a metric space (not necessarily compact) and let A be a set of real-valued functions with domain M. Let $a \in M$. The set A is said to be *equicontinuous at the point a* if the following condition is satisfied: for all $\varepsilon > 0$ there exists $\delta > 0$, such that for all $f \in A$ we have $|f(x) - f(a)| < \varepsilon$ for all $x \in M$ that satisfy $d(x, a) < \delta$. The set A is said to be *equicontinuous* if it is equicontinuous at all points $a \in M$.

The condition that the set A of functions is equicontinuous at the point $a \in M$ needs four quantifiers:

$$(\forall \varepsilon > 0)(\exists \delta > 0)(\forall f \in A)(\forall x \in M)\big(d(x, a) < \delta \Rightarrow |f(x) - f(a)| < \varepsilon\big)$$

If M is compact, we have a nice symmetrical description of equicontinuity analogous, in both statement and proof, to proposition 4.11.

Proposition 4.25 *Let M be compact and let $A \subset C(M)$ be equicontinuous. Then for all $\varepsilon > 0$ there exists $\delta > 0$, such that for all $f \in A$ we have $|f(x) - f(y)| < \varepsilon$ for all x and y in M that satisfy $d(x, y) < \delta$.*

Proof Let $\varepsilon > 0$. For each $a \in M$ there exists $\rho_a > 0$, such that for all x that satisfy $d(x, a) < \rho_a$ and for all $f \in A$ we have $|f(x) - f(a)| < \frac{1}{2}\varepsilon$. Since M is compact there exist finitely many points $a_1, a_2, \ldots, a_n$ in M, such that the balls

$$B_{\rho_{a_k}/2}(a_k), \quad (k = 1, 2, \ldots, n),$$

cover M. Let

$$\delta = \frac{1}{2} \min_{1 \leq k \leq n} (\rho_{a_k}).$$

Now let $f \in A$ and suppose that x and y in M satisfy $d(x, y) < \delta$. There exists k such that $x \in B_{\rho_{a_k}/2}(a_k)$. Since $\delta \leq \frac{1}{2}\rho_{a_k}$ we have $d(x, y) < \frac{1}{2}\rho_{a_k}$, so that both x and y lie in the ball $B_{\rho_{a_k}}(a_k)$. Hence $|f(x) - f(a_k)| < \frac{1}{2}\varepsilon$ and $|f(y) - f(a_k)| < \frac{1}{2}\varepsilon$. Therefore $|f(x) - f(y)| < \varepsilon$, as required. □

The notion of equicontinuity is commonly applied to families of functions. The meaning should be obvious: the family $(f_i)_{i \in I}$ is equicontinuous when its set of terms $\{f_i : i \in I\}$ is. Most often the family is a sequence. We can extract information about the pointwise convergence of an equicontinuous sequence $(f_n)_{n=1}^{\infty}$ of functions on a metric space M, from its pointwise convergence on a dense subset of M. A subset D of M is said to be dense if its closure $\overline{D}$ is all of M. If the space M is compact we obtain in addition uniform convergence of the sequence $(f_n)_{n=1}^{\infty}$.

Proposition 4.26 *Let M be a metric space (not necessarily compact), let $(f_n)_{n=1}^{\infty}$ be an equicontinuous sequence of real-valued functions with domain M and suppose that it is pointwise convergent at each point of a set D that is dense in M. Then the sequence $(f_n)_{n=1}^{\infty}$ is pointwise convergent in M.*

Proof Let $x \in M$. We shall show that $(f_n(x))_{n=1}^{\infty}$ is a Cauchy sequence of real numbers.

Let $\varepsilon > 0$. By equicontinuity, and because $\overline{D} = M$, there exists $a \in D$, such that $|f_k(x) - f_k(a)| < \varepsilon/3$ for all k. Because the sequence $(f_n(a))_{n=1}^{\infty}$ converges, there exists N, such that $|f_m(a) - f_n(a)| < \varepsilon/3$ for all $m \geq N$ and $n \geq N$. For such m

and n we must therefore have $|f_m(x) - f_n(x)| < \varepsilon$. We conclude that $(f_n(x))_{n=1}^{\infty}$ is a Cauchy sequence of real numbers; hence convergent. □

Proposition 4.27 *Let* M *be compact and let* $(f_n)_{n=1}^{\infty}$ *be an equicontinuous sequence in* $C(M)$ *that is also pointwise convergent. Then the sequence* $(f_n)_{n=1}^{\infty}$ *is convergent in* $C(M)$, *that is, it converges uniformly to a continuous function.*

Proof We show that the sequence $(f_n)_{n=1}^{\infty}$ converges uniformly to its pointwise limit g.

Let $\varepsilon > 0$. By equicontinuity, for each $a \in M$ there exists $\rho_a > 0$, such that for all n we have $|f_n(x) - f_n(a)| < \varepsilon/3$ for all x that satisfy $d(x, a) < \rho_a$. By the compactness of M, there exist finitely many points $a_1, a_2, \ldots, a_p$, such that the balls $B_{\rho_{a_k}}(a_k)$, $(k = 1, 2, \ldots, p)$, cover M. By the pointwise convergence of f_n to g, there exists N, such that $|f_n(a_k) - g(a_k)| < \varepsilon/3$ for all $n \geq N$ and for each of the finitely many indices k.

With ε as given in the previous paragraph we let $x \in M$. There exists an index k in the range $1 \leq k \leq p$, such that $d(x, a_k) < \rho_{a_k}$. Hence for all n we have $|f_n(x) - f_n(a_k)| < \varepsilon/3$, and letting $n \to \infty$ we find $|g(x) - g(a_k)| \leq \varepsilon/3$.

With ε and x as given we let $n \geq N$. The number N was defined in the previous paragraph but one and depends only on ε, in particular it is independent of x, as the reader should check at this point. With k as defined in the previous paragraph (k depends only on ε and x, which, again, the reader might check), the three inequalities

$$|g(x) - g(a_k)| \leq \varepsilon/3, \quad |f_n(a_k) - g(a_k)| < \varepsilon/3, \quad |f_n(x) - f_n(a_k)| < \varepsilon/3$$

are all valid and imply $|f_n(x) - g(x)| < \varepsilon$. We conclude that $f_n \to g$ uniformly. □

Proposition 4.28 (Ascoli's Theorem[1]) *Let* M *be compact and let* $A \subset C(M)$. *Then for the set* A *to be relatively compact in* $C(M)$ *it is necessary and sufficient that it is bounded and equicontinuous.*

Proof of Sufficiency Assume that A is bounded and equicontinuous. Let $(f_n)_{n=1}^{\infty}$ be a sequence in A. It is enough to show that the sequence $(f_n)_{n=1}^{\infty}$ has a subsequence convergent in $C(M)$.

Now M being compact, it has a countable dense subset D, that is, there exists a countable set D such that $\overline{D} = M$ (the reader is asked to prove this; see exercise 3). By propositions 4.26 and 4.27, it is enough to show that $(f_n)_{n=1}^{\infty}$ has a subsequence that is pointwise convergent in D. Because A is bounded, there exists an interval $J = [-K, K]$, such that the values of every function in A lie in J. The restriction of f to D can be viewed as an element of the countable product J^D. The required conclusion follows since the space J^D is compact (see 4.3 exercises 10 and 14).

---

[1] Often called the Arzela-Ascoli theorem. It seems that Arzela proved that the condition is necessary, shortly after Ascoli proved that it is sufficient.

Proof of Necessity Assume that A is relatively compact. It is bounded since all compact sets are bounded. Let $a \in M$, suppose that A is not equicontinuous at a and derive a contradiction. By this assumption, there exists $\varepsilon > 0$, together with sequences $(f_n)_{n=1}^{\infty}$ in A, and $(x_n)_{n=1}^{\infty}$ in M, such that $x_n \to a$ and $|f_n(x_n) - f_n(a)| \geq \varepsilon$ for all n. Since A is relatively compact, we may assume that the sequence $(f_n)_{n=1}^{\infty}$ is uniformly convergent, with limit $g \in C(M)$, say. Now we have

$$0 < \varepsilon \leq |f_n(x_n) - f_n(a)|$$
$$\leq |f_n(x_n) - g(x_n)| + |g(x_n) - g(a)| + |g(a) - f_n(a)|$$

and all three terms of the right-hand member tend to 0 as $n \to \infty$; a plain contradiction.   □

Another proof of necessity is suggested in exercise 6. It is a direct proof, not proceeding by contradiction, and neither deploying sequences nor open covers.

In most cases when compactness is required of a subset of a function space, such as is so often the case in the study of partial differential equations, Ascoli's theorem is at work behind the scenes. It is therefore hard to overestimate its importance.

4.5.1 Peano's Existence Theorem

We shall describe a simple application of Ascoli's theorem to proving the existence of a local solution to the initial value problem for an ordinary differential equation. The problem is usually posed in the form

$$y' = f(x, y), \quad y(x_0) = y_0 \tag{4.1}$$

where $f : A \to \mathbb{R}$, and A is an open set in $\mathbb{R}^2$. The numbers x_0 and y_0 are the coordinates of a point in A. By a solution to the problem is usually meant a function

$$\phi :]x_0 - h, x_0 + h[\to \mathbb{R},$$

whose graph lies in A, satisfies $\phi(x_0) = y_0$, and for all x in $]x_0 - h, x_0 + h[$ satisfies

$$\phi'(x) = f(x, \phi(x)).$$

The solution is called local because h is unknown in advance.

Peano's existence theorem asserts that if f is continuous a local solution exists for some $h > 0$. In fact, suppose that for $|x - x_0| \leq a$ and $|y - y_0| \leq b$ we have the bound $|f(x, y)| \leq M$. Then a solution exists for

$$h = \min\left(a, \frac{b}{M}\right).$$

To simplify the notation, let us suppose that $x_0 = y_0 = 0$. There is really no loss of generality since this can be accomplished by a translation of the coordinates. Now we will show that a continuous function ϕ exists on the interval $[-h, h]$ that satisfies

$$\phi(x) = \int_0^x f(t, \phi(t)) \, dt, \quad (-h \leq x \leq h). \tag{4.2}$$

By the fundamental theorem of calculus ϕ is a solution of the initial value problem in the open interval $]-h, h[$.

We shall prove that a solution of (4.2) exists to the right of $x = 0$; more precisely, in the interval $[0, h]$. The plan is to construct a sequence of functions $(\psi_n)_{n=1}^\infty$ in $C[0, h]$ that might reasonably be expected to approximate a solution to the right of $x = 0$, building ψ_n as a piece-wise affine function in steps. The approximations, though continuous, will fail to be differentiable at finitely many points. It may seem odd to use, as approximations to a solution of a differential equation, functions that lack in part derivatives. Yet this turns out to be a very fruitful and practical approach both for ordinary differential equations and for partial differential equations.

For each n we partition the interval $[0, h]$ by the $n + 1$ points

$$t_k^{(n)} = \frac{kh}{n}, \quad (k = 0, 1, \ldots, n).$$

Fixing n, we define ψ_n in the subintervals of the partition, inductively from left to right, as follows. In the first subinterval of the partition we set

$$\psi_n(x) = f(0, 0)x, \quad (0 \leq x \leq t_1^{(n)}).$$

When, for a given n, we have defined $\psi_n(x)$ in k successive subintervals up to $x = t_k^{(n)}$, we set

$$\psi_n(x) = \psi_n(t_k^{(n)}) + f(t_k^{(n)}, \psi_n(t_k^{(n)}))(x - t_k^{(n)}), \quad (t_k^{(n)} < x \leq t_{k+1}^{(n)})$$

thus extending $\psi_n(x)$ up to $x = t_{k+1}^{(n)}$.

Obviously the function ψ_n is continuous and piece-wise affine; it has constant slope on each subinterval $]t_k^{(n)}, t_{k+1}^{(n)}[$, and its right derivative at the left endpoint would be "correct" for a solution, were it not for the fact that the ordinate is presumably not quite correct, except, of course, at $x = 0$. We note also that it satisfies the estimate:

$$|\psi_n(x) - \psi_n(t_k^{(n)})| \leq \frac{Mh}{n}, \quad (t_k^{(n)} < x \leq t_{k+1}^{(n)}). \tag{4.3}$$

This method of approximating the solution is associated with Euler and most modern methods in use are simply improvements on it. We are not concerned here with proving that the approximations converge to a solution, but only with using them to prove that a solution exists.

An important point, which explains the definition of h, is that the graph of the approximation ψ_n is trapped in the rectangle defined by the inequalities $|x| \leq a$, $|y| \leq b$.

Exercise Prove this. A hint: every segment of the graph of ψ_n has slope less than M.

Having defined the approximate solutions we look at the error. More precisely, since we do not yet have a solution we can compare them to, we set

$$E_n(x) := \psi_n(x) - \int_0^x f(t, \psi_n(t)) \, dt, \quad (0 \leq x \leq h)$$

By the fundamental theorem of calculus, valid here because ψ_n, though lacking derivatives at finitely many points, is continuous, and its derivative piece-wise continuous, we have:

$$E_n(x) = \int_0^x \left(\psi_n'(t) - f(t, \psi_n(t)) \right) dt.$$

Hence

$$|E_n(x)| \leq \int_0^h \left| \psi_n'(t) - f(t, \psi_n(t)) \right| dt, \quad (0 \leq x \leq h)$$

We also have

$$\int_0^h \left| \psi_n'(t) - f(t, \psi_n(t)) \right| dt = \sum_{k=0}^{n-1} \int_{t_k^{(n)}}^{t_{k+1}^{(n)}} \left| \psi_n'(t) - f(t, \psi_n(t)) \right| dt$$

$$= \sum_{k=0}^{n-1} \int_{t_k^{(n)}}^{t_{k+1}^{(n)}} \left| f(t_k^{(n)}, \psi_n(t_k^{(n)})) - f(t, \psi_n(t)) \right| dt$$

and therefore, for $0 \leq x \leq h$ we have the estimate

$$|E_n(x)| \leq \sum_{k=0}^{n-1} \int_{t_k^{(n)}}^{t_{k+1}^{(n)}} \left| f(t_k^{(n)}, \psi_n(t_k^{(n)})) - f(t, \psi_n(t)) \right| dt.$$

Let $\varepsilon > 0$. Since f is uniformly continuous on compact sets in $\mathbb{R}^2$ (proposition 4.11), in particular in the rectangle defined by $|x| \leq h$, $|y| \leq b$, and in view of the estimate (4.3), we can find N, such that for $n \geq N$ we have

$$\left| f(t_k^{(n)}, \psi_n(t_k^{(n)})) - f(t, \psi_n(t)) \right| < \varepsilon,$$

for $t_k^{(n)} < t < t_{k+1}^{(n)}$, and $k = 0, 1, \ldots, n - 1$. Therefore, for $n \geq N$ we have

$$|E_n(x)| < h\varepsilon, \quad (0 \leq x \leq h).$$

It follows that $E_n(x) \to 0$ uniformly for $x \in [0, h]$.

After all this preparation we are ready to call in Ascoli's theorem. The sequence $(\psi_n)_{n=1}^{\infty}$ is bounded by b in absolute value, and it is equicontinuous, since for all n we have $|\psi_n'(x)| \leq M$ except at a finite number of points x in $[0, h]$ (a conclusion the reader is asked to consider; see exercise 1). Therefore, by Ascoli's theorem, there exists a uniformly convergent subsequence, say, $\psi_{k_n} \to \phi$. Again, because f is uniformly continuous in the rectangle $|x| \leq a, |y| \leq b$, we find that

$$f\big(x, \psi_{k_n}(x)\big) \to f\big(x, \phi(x)\big)$$

uniformly on the interval $0 \leq x \leq h$. We can therefore invert the integral and limit and deduce

$$\lim_{n \to \infty} E_{k_n}(x) = \lim_{n \to \infty} \psi_{k_n}(x) - \lim_{n \to \infty} \int_0^x f(t, \psi_{k_n}(t)) \, dt$$

$$= \phi(x) - \int_0^x f\big(t, \phi(t)\big) \, dt$$

Since the limit is known to be 0 we finally obtain

$$\phi(x) = \int_0^x f\big(t, \phi(t)\big) \, dt.$$

A local solution therefore exists to the initial value problem. Notice that we have not proved that the approximations converge; only that a subsequence of them converges. It remains to dispose of the solution to the left of $x = 0$. Either we can handle it by a similar method (constructing Euler approximations in steps going to the left), or else, we can reflect the coordinates about the y-axis, point out that the reflected problem has a solution to the right of $x = 0$, which, on reflection, converts to a solution of the original problem to the left of $x = 0$.

4.5.2 Exercises

1. a) Let $A \subset C[a, b]$, and suppose that every f in A is differentiable in the open interval $]a, b[$. Suppose that there exists K, such that for all $f \in A$ and $x \in]a, b[$ we have $|f'(x)| \leq K$. Show that A is equicontinuous.
   b) More generally, we can assume that for each $f \in A$ there exists a finite set $D_f \subset]a, b[$, such that $|f'(x)| \leq K$ for all $x \in]a, b[\setminus D_f$ (we don't even

need f to be differentiable at points in D_f), and obtain the same conclusion. Compare the proof of Peano's theorem, in which the approximating functions are piece-wise affine.

2. a) Suppose that the Euler approximations in the proof of Peano's theorem converge pointwise in the interval $[0, h]$ to a function ϕ. Show that ϕ is a solution.

   b) Suppose that the initial value problem has a unique solution ϕ in the interval $[0, h]$. Show that the Euler approximations converge uniformly to ϕ. They therefore furnish a practical method for approximating the solution, if the latter is known to be unique.

   Note. Practical criteria are known that guarantee uniqueness; see 6.5.

3. Supply a step in the proof of Ascoli's theorem. Show that a compact space X has a countable dense subset. That is, there exists a countable set D such that $\overline{D} = X$.

   Hint. For every n you can cover X with finitely many open balls of radius $1/n$.

4. Let M be compact and let $A \subset C(M)$. Assume that A is equicontinuous, and that for each $x \in M$ the set $\{f(x) : f \in A\}$ is bounded. Show that A is bounded.

5. Let M be compact. Prove the following exhaustive description of the compact sets in $C(M)$: they are those that are closed, bounded and equicontinuous.

6. Let M be compact and let $v : C(M) \times M \to \mathbb{R}$ be the *evaluation mapping* $v(f, x) = f(x)$.

   a) Show that v is continuous.

   b) Let $A \subset C(M)$ and let $\varepsilon > 0$. Show that the set

   $$\left\{ ((f, x), (g, y)) \in (A \times M) \times (A \times M) : |f(x) - g(y)| < \varepsilon \right\}$$

   is open in the product space $(A \times M) \times (A \times M)$ and includes the diagonal (4.3 exercise 13).

   c) Assume that A is compact and deduce the necessity part of Ascoli's theorem from the previous item and 4.3 exercise 13.

7. Let $J : C^1[a, b] \to C[a, b]$ (see 4.3 exercise 2 for the definition of $C^1[a, b]$) be the identity injection, $J(f) = f$. Show that J maps bounded sets in $C^1[a, b]$ on to relatively compact sets in $C[a, b]$.

   Note. This is probably the simplest example of what is often called a *compact embedding* of function spaces. They are of crucial importance in the theory of partial differential equations and generally involve two function spaces, the first of which is a linear subspace of the second, whilst functions of the first have more derivatives than functions of the second. We have discussed elsewhere the notion of embedding, for example in 3.2 exercise 1 and in section 2.5.

8. Let $K(x, y)$ be a continuous function of two real variables in the range $a \leq x, y \leq b$. For each $u \in C[a, b]$ we define

$$(Tu)(x) = \int_a^b K(x, y)u(y)\, dy, \quad (a \leq x \leq b).$$

Show that this defines a linear mapping $T : C[a, b] \to C[a, b]$ and that T maps bounded sets on to relatively compact sets.

Hint. Show that the image of the unit ball in $C[a, b]$ is an equicontinuous set of functions on $[a, b]$.

Note. A linear mapping $T : X \to X$, where X is a Banach space, with the property that T maps bounded sets on to relatively compact sets, is called a compact (linear) mapping. They are important in functional analysis because their *spectrum* has a simple structure. They have many applications to differential equations, both ordinary and partial. In connection with this exercise see also the discussion of the Fredholm integral equation (4.5.1).

9. This exercise assumes some knowledge of complex analysis. The metric space $O(A)$ of complex analytic functions in an open set $A \subset \mathbb{C}$ was described in section 4.3, exercises 27 and 28, and the preamble to them.

   a) Let $(f_n)_{n=1}^\infty$ be a sequence in $O(A)$ such that the functions f_n are uniformly bounded on each compact set $K \subset A$. Precisely, this means that for each compact $K \subset A$ there exists M, such that $|f_n(x)| \leq M$ for all n and all $x \in K$. Show that the sequence $(f_n)_{n=1}^\infty$ has a subsequence convergent in $O(A)$.

   Hint. Show that the sequence $(f_n)_{n=1}^\infty$ is equicontinuous in A. Given $z_0 \in A$ take a circle centred at z_0, lying, along with its interior disc, in A. Estimate $f_n(z) - f_n(z_0)$ by the Cauchy integral formula.

   b) Show that the relatively compact sets in $O(A)$ are precisely those that are bounded in the supremum norm on each compact subset of A.

   Note. The result is called Montel's theorem. It seems to show that something close to the Heine-Borel theorem is true for $O(A)$, which is, of course, not a normed space, if by a bounded set of functions we mean one that is bounded in $C(K)$ for each compact set $K \subset A$.

4.5.3   Pointers to Further Study

→   Ordinary differential equations (e.g. Sturm-Liouville problems).
→   Partial differential equations (e.g. existence of solutions).
→   Complex analysis.
→   Functional analysis (e.g. spectral theory)

Chapter 5
Separable Spaces

> *And further, by these, my son, be admonished: of making many books there is no end; and much study is a weariness of the flesh.*
>
> *Ecclesiastes. Author unknown.*

We introduce a topological property of metric spaces with strong practical implications. Upon it depends digitalisation, which pervades almost all of present technology. For this chapter, a good understanding of the concept of countable set is needed.

5.1 Dense Subsets of a Metric Space

The notion of a dense subset of a metric space has been touched on in passing several times, notably in section 4.5 where it even played a key role, but it has not so far received a highlighted definition. Now we begin a more detailed study of the concept, which will be continued in the next chapter.

Definition A set A in a metric space (X, d) is said to be dense in X if its closure is X, in symbols $\overline{A} = X$.

Like any other important concept of metric space theory, density of A in X may be viewed in a number of different ways. For example:

a) Each $x \in X$ is the limit of some sequence $(a_n)_{n=1}^{\infty}$ in A.
b) Every point in X that is not in A is a limit point of A.
c) Every non-empty open set in X contains a point of A.

Let X and Y be metric spaces and let $f : X \to Y$ be continuous. If A is dense in X then f is completely determined by its restriction to A. For if $x \in X$ there exists a sequence $(a_n)_{n=1}^{\infty}$ in A, such that $a_n \to x$. Then $f(a_n) \to f(x)$.

© The Author(s), under exclusive license to Springer Nature Switzerland AG 2022
R. Magnus, *Metric Spaces*, Springer Undergraduate Mathematics Series,
https://doi.org/10.1007/978-3-030-94946-4_5

It is, however, not always the case that a continuous mapping, $f : A \to Y$, defined initially on A, has a continuous extension to X, though extensions, possibly discontinuous, obviously exist. The following result is therefore often useful and shows a benefit arising from having uniform continuity and completeness acting in tandem.

Proposition 5.1 *Let X and Y be metric spaces, let A be a dense set in X and assume that Y is complete. Let $f : A \to Y$ be uniformly continuous. Then f has a unique continuous extension $F : X \to Y$. The extension F is also uniformly continuous.*

Proof Let $x \in X$. We first show how to define F at x. Since A is dense in X, there exists a sequence $(a_n)_{n=1}^{\infty}$ in A, such that $a_n \to x$. Let $\varepsilon > 0$. Since f is uniformly continuous in A, there exists $\delta > 0$, such that $d(f(x), f(x')) < \varepsilon$ for all x and x' in A that satisfy $d(x, x') < \delta$. Since the sequence $(a_n)_{n=1}^{\infty}$ is convergent it is also a Cauchy sequence. Hence there exists N, such that $d(a_n, a_m) < \delta$ for all $n \geq N$ and $m \geq N$. For such n and m we have $d(f(a_m), f(a_n)) < \varepsilon$. We conclude that $(f(a_n))_{n=1}^{\infty}$ is a Cauchy sequence in Y, and since Y is complete it is convergent.

Obviously we must define $F(x)$ to be $\lim_{n \to \infty} f(a_n)$, but we need to show that the limit only depends on x and not on the way we chose a sequence with limit x. In other words we need to show that the limit is the same if we replace the sequence $(a_n)_{n=1}^{\infty}$ by another sequence $(b_n)_{n=1}^{\infty}$ in A that also converges to x.

Here a trick saves the day. Build a third sequence $(c_n)_{n=1}^{\infty}$ in A, also converging to x, by using the terms of $(a_n)_{n=1}^{\infty}$ and $(b_n)_{n=1}^{\infty}$ alternately. Now $\lim_{n \to \infty} f(c_n)$ exists by the arguments given before, and hence the limits $\lim_{n \to \infty} f(a_n)$ and $\lim_{n \to \infty} f(b_n)$ must be the same. The mapping $F : X \to Y$ is therefore well-defined.

It is clear that F has to be defined in this way if it is to be a continuous extension; so a continuous extension of f, if one exists, is uniquely defined.

Finally we show that F is uniformly continuous. Let $\varepsilon > 0$. Since f is uniformly continuous on A there exists $\delta > 0$, such that $d(f(x), f(x')) < \varepsilon$ for all x and x' in A that satisfy $d(x, x') < \delta$. Let c and c' be points in X that satisfy $d(c, c') < \delta/2$. There exist sequences $(a_n)_{n=1}^{\infty}$ and $(a_n')_{n=1}^{\infty}$ in A that converge to c and c' respectively. Since $d(c, c') < \delta/2$ and the metric d is continuous on the product space X^2 we can conclude that there exists N, such that for $n \geq N$ we have $d(a_n, a_n') < \delta$. Therefore for $n \geq N$ we also have $d(f(a_n), f(a_n')) < \varepsilon$. Now $f(a_n) \to F(c)$ and $f(a_n') \to F(c')$. We conclude, again using the continuity of d, that $d(F(c), F(c')) \leq \varepsilon$. Summarising what we have here: if c and c' are points in X that satisfy $d(c, c') < \delta/2$, then $d(F(c), F(c')) \leq \varepsilon$. But this says that F is uniformly continuous.   □

Definition Let (X, d) and (Y, d') be metric spaces and let $f : X \to Y$. The mapping f is called an *isometry of X on to Y* if it is a bijection and satisfies $d'(f(a), f(b)) = d(a, b)$ for all a and b in X.

In other words an isometry is a bijection that preserves the metric. Obviously an isometry is a special kind of homeomorphism that makes X and Y, which may differ wildly in their construction, "the same", not just topologically but in their entire metric space structure.

Mappings that are injective, preserve the metric, but are not surjective, are sometimes called isometries, though properly, if we need to avoid confusion, we should say that such a mapping is an isometry from its domain on to its range; or that f is an isometry on to $f(X)$. Isometries between normed spaces were the subject of section 2.7.

Proposition 5.2 *Let X and Y be metric spaces, let $A \subset X$ and assume that A is dense in X. Assume that Y is complete. Let $f : A \to Y$ be an isometry of A on to $f(A)$. Then the unique continuous extension F of f to X is an isometry of X on to $F(X)$.*

Proof Let $c, c' \in X$ and let $a_n, a'_n \in A$ with $a_n \to c$ and $a'_n \to c'$. Then, by the continuity of d, we have

$$d(F(c), F(c')) = \lim_{n \to \infty} d(f(a_n), f(a'_n)) = \lim_{n \to \infty} d(a_n, a'_n) = d(c, c').$$

$\square$

Propositions 5.1 and 5.2 have numerous applications in functional analysis. A typical situation is that X and Y are Banach spaces (and therefore Y complete), A is a linear subspace of X that is dense in X, and $f : A \to Y$ is linear and continuous. Then f is also Lipschitz continuous, as are all continuous, linear mappings of normed spaces, and so proposition 5.1 guarantees an extension of f to all of X. Moreover, the extension of f is also linear and has the same Lipschitz constant (see exercise 2).

The point of such applications is often to define f in all of X in a situation when f is easy to define in A because the elements of A are in some sense especially simple. It is important to know a range of dense subspaces of the important spaces of functional analysis, that can serve this purpose. Spaces that build on spaces of integrable functions are especially important, and we shall look at some examples in the exercises to this section.

5.1.1 Defining a Vector-Valued Integral

We shall venture a little into the realm of analysis to give a simple, but useful, application of proposition 5.1.

Let E be a Banach space (which, we recall, means a complete normed space). We wish to define a simple notion of integral $\int_a^b f$ for functions $f : [a, b] \to E$. In the case that $E = \mathbb{R}^n$, the realm of multivariate analysis, this can be accomplished by considering the coordinates of f, which are real valued functions and can be integrated in the normal way and assembled into the vector $\int_a^b f$. To be precise, for $f = (f_1, \ldots, f_n)$ we can define

$$\int_a^b f = \left(\int_a^b f_1, \ldots, \int_a^b f_n \right)$$

provided the coordinate functions are integrable. For a Banach space E in all generality a coordinate approach is not available and we need to find an alternative.

Let $B([a, b], E)$ be the vector space of all bounded functions from $[a, b]$ to E. We can use the norm

$$\|f\|_\infty = \sup_{a \leq t \leq b} \|f(t)\|$$

on $B([a, b], E)$. There is little chance of being able to define an integral for all functions in $B([a, b], E)$, but we can start by defining the integral for a "small" subspace of it, consisting of functions with a simple structure.

We say that $f : [a, b] \to E$ is a *step function* if there exists a partition $(t_0, t_1, \ldots, t_m)$ of $[a, b]$ and vectors $c_0, \ldots, c_{m-1}$ in E, such that

$$f(s) = c_k, \quad (t_k < s < t_{k+1}), \quad (k = 0, 1, \ldots m - 1).$$

We will occasionally need step functions defined in an unbounded interval I, for example in $[0, \infty[$ or for all of $\mathbb{R}$. A function $f : I \to \mathbb{R}$ is called a step function, if its restriction to some bounded interval $[a, b] \subset I$ is a step function in the sense defined above, and it is zero in the complement of $[a, b]$ in I. We do not allow a step function to have infinitely many steps.

Let A be the space of all step functions with domain $[a, b]$. There is a detail that we shall leave to the reader as an exercise: one needs to show that A is a linear subspace of $B([a, b], E)$. Then one defines the integral of a step function as a mapping $J : A \to E$ given by

$$J(f) = \sum_{k=0}^{m-1} (t_{k+1} - t_k) c_k$$

where f is the step function defined above.

The sequences of points $(t_k)_{k=0}^m$ and vectors $(c_k)_{k=0}^{m-1}$ are those associated with the structure of f as a step function. This indicates another difficulty to be overcome, that these sequences are not uniquely defined by f; for example one can always introduce additional partition points. So one must show that the function J is well-defined, and is, moreover, a linear mapping from the space of step functions to E. The details are left to the reader, but there is a key idea, the refinement of a partition, that should be familiar from the elementary Riemann-Darboux integral (see exercise 4).

Assuming the difficulties disposed of, we easily obtain the inequality

$$\|J(f)\| \le (b-a)\|f\|_\infty.$$

Hence J is continuous, indeed Lipschitz continuous. By the extension rule there exists a unique continuous extension of J to the closure $\overline{A}$ of A in $B([a,b], E)$. The extension is also linear (exercise 2). We shall use the traditional notation

$$\int_a^b f \quad \text{or} \quad \int_a^b f(x)\,dx$$

for this extension (the second is sometimes more convenient, especially when the integrand involves standard functions of calculus, such as e^x). In addition to linearity we also have the same Lipschitz constant as for J (exercise 2), expressed by the inequality:

$$\left\| \int_a^b f \right\| \le (b-a)\|f\|_\infty.$$

In a full treatment of the integral one would go on to prove the familiar properties, adapted to the vector-valued context. These include the inequality

$$\left\| \int_a^b f \right\| \le \int_a^b \|f\|,$$

where $\|f\|$ denotes the function $\|f(x)\|$, not to be confused with the function norm $\|f\|_\infty$, the rule for the join of intervals

$$\int_a^b f + \int_b^c f = \int_a^c f,$$

and a version of the fundamental theorem of calculus.

There remain some big questions that the reader has perhaps spotted. What exactly is the space $\overline{A}$? It may be described as the space of all functions $f : [a,b] \to E$ that may be uniformly approximated by step functions. This is not much clearer. Does it include all continuous functions? What discontinuous functions does it include (apart from step functions)? In the case that $E = \mathbb{R}$ does it include all Riemann integrable functions?

In fact we introduced this space in 4.3 exercise 25 and its preamble, to which we refer the reader. It is the class of regulated functions, which N. Bourbaki promoted as the correct domain in which to define an elementary type of integral, that fell short of the Riemann integral and the more sophisticated Lebesgue integral, and was useful for applications (it includes the sort of discontinuities—namely jumps—that are likely to occur in applications to technology), whilst disparaging the Riemann integral as only a "mildly interesting exercise".[1]

Another big question, common to all notions of an integral, is that it is one thing to define the concept of an integral, but quite another thing to calculate one. This is, of course, a major goal of calculus. Some of the gaps in the above discussion are topics in the following exercise set. The version of the vector valued integral just introduced, is also called the *regulated integral*.

Step functions have an obvious relevance to Riemann-Darboux integrals, but less because of uniform approximation by step functions, and more because of *approximation in the mean*. It follows immediately by the definition of the Riemann-Darboux integral of a bounded function $f : [a, b] \to \mathbb{R}$, that for every $\varepsilon > 0$ there exists a step function g, such that

$$\int_a^b |f - g| < \varepsilon.$$

Here the approximation uses the norm (more precisely this is a seminorm but this can be dealt with; see 4.2 exercise 1.2.6)

$$\|f\|_1 = \int_a^b |f|.$$

To describe the closure of the space of step functions with respect to this norm we must extend the integral to the Lebesgue integral. The space $L^1[a, b]$ consists of all Lebesgue integrable functions $f : [a, b] \to \mathbb{R}$. It is virtually the *raison d'etre* of the Lebesgue integral that this space is complete. Moreover the space of step functions is dense in it.

5.1.2   Exercises

1. In the context of proposition 5.1, if the mapping f satisfies a Lipschitz condition, then so does F. Suppose that

$$\|f(x) - f(y)\| \le K \|x - y\|, \quad (x, y \in A).$$

Show that

$$\|F(x) - F(y)\| \le K \|x - y\|, \quad (x, y \in X).$$

---

[1] Quoted from J. Dieudonné, Foundations of Modern Analysis, chapter 8.

2. In the context of proposition 5.1, suppose that X and Y are Banach spaces and
   that A is a dense vector subspace of X. Suppose further that $f : A \to Y$ is linear
   and continuous. Prove that the unique continuous extension F to X is linear.
   Prove, moreover, that if the linear mapping f satisfies

$$\|f(x)\| \leq K \|x\|$$

   for all $x \in A$, then the extension F satisfies

$$\|F(x)\| \leq K \|x\|$$

   for all $x \in X$. Deduce that f and F have the same operator norm.
3. Let α and β be non-zero real numbers. Define the set $M \subset \mathbb{R}$ by

$$M := \{m\alpha + n\beta : (m, n) \in \mathbb{Z}^2\}.$$

We distinguish two cases:

Case A. M has a lowest positive element.
Case B. M does not have a lowest positive element.

a) Show that case A occurs if and only if α/β is rational; and if this is the case
   then M consists of all integral multiples of its lowest positive element.
b) In case B (when α/β is irrational) the set M is dense in $\mathbb{R}$.
c) If β is irrational the line $y = \beta x$ in the coordinate plane $\mathbb{R}^2$ passes arbitrarily
   close to the lattice $\mathbb{Z}^2$ of points with integral coordinates.

4. Given two partitions

$$a = t_0 < t_1 < \cdots < t_m = b, \quad a = s_0 < s_1 < \cdots < s_\ell = b$$

of the interval $[a, b]$ one can produce a new partition, finer than both, by uniting
their points. So given two step functions f and g, there is a partition of $[a, b]$,
in the (open) subintervals of which both functions are constant. This idea can
be used to check the points raised in the definition of the vector valued integral,
which involve the definition of the integral for step functions, and, in particular,
whether the definition even makes sense.

Let E be a Banach space. Show the following:

a) The set of all step functions $f : [a, b] \to E$ is a linear subspace of the space
   $B([a, b], E)$ of bounded functions.
b) The integral $J(f)$ of the step function f is well-defined, that is, it is
   independent of the partition, in the (open) subintervals of which f is constant.
c) The integral in the space of step functions is a linear mapping with
   codomain E.

d) The integral in the space of step functions satisfies

$$\|J(f)\| \le (b-a)\|f\|_\infty.$$

5. Prove the inequality

$$\left\| \int_a^b f \right\| \le \int_a^b \|f\|,$$

for the regulated integral.

Hint. Prove it first for step functions.

6. In this exercise we assume a slight acquaintance with the Lebesgue integral of functions of one real variable. The (complex) space $L^1(\mathbb{R})$ consists of all Lebesgue integrable functions $f : \mathbb{R} \to \mathbb{C}$, with the norm

$$\|f\|_1 = \int_{-\infty}^\infty |f|.$$

This is a natural domain for the Fourier transform, which we define for each $f \in L^1(\mathbb{R})$ to be the function

$$\hat{f}(\xi) = \int_{-\infty}^\infty e^{-ix\xi} f(x)\, dx.$$

Readers with a minimal knowledge of the Lebesgue integral should recognise that the integral on the right exists and that we have the estimate

$$|\hat{f}(\xi)| \le \|f\|_1, \quad (\xi \in \mathbb{R})$$

The Fourier transform of f is therefore a bounded function and the transform operator, $f \mapsto \hat{f}$, maps $L^1(\mathbb{R})$ into the (complex) space of bounded functions $B(\mathbb{R})$, is linear and satisfies

$$\|\hat{f}\|_\infty \le \|f\|_1.$$

The mapping $f \mapsto \hat{f}$ is therefore also continuous from $L^1(\mathbb{R})$ to $B(\mathbb{R})$.

Now use the fact that the space of all step functions $g : \mathbb{R} \to \mathbb{C}$ is dense in $L^1(\mathbb{R})$ to prove that $\hat{f} \in C_0(\mathbb{R})$ (see 2.2 exercise 19). Recall that a step function with domain $\mathbb{R}$ is intended to be zero outside some bounded interval.

5.2 Separability

The eponymous topic of this chapter is a property that was mentioned in the brief list of topological properties in section 2.6. It cropped up in the proof of Ascoli's theorem in 4.5, though it remained unnamed.

Definition A metric space (X, d) is said to be separable if it includes a countable dense subset.

Two familiar examples of a separable space are:

a) The real line $\mathbb{R}$ includes the countable dense subset $\mathbb{Q}$.
b) The coordinate space $\mathbb{R}^n$ includes the countable dense subset $\mathbb{Q}^n$.

Separability is of immense practical importance that pervades almost the whole of modern technology.[2] The allusion is to digitalisation. A computer can only represent internally a finite number of things. Yet it must handle things like real numbers, curves, pictures of physical objects etc. These are all usually modelled as elements of infinite sets. Yet all of these need to be approximated by objects that are selected from a finite list.

Suppose that a class of objects that we wish to represent can be thought of as a separable metric space X, which therefore has a countable dense subset A. All points in X can be approximated with arbitrary accuracy by points drawn from A. Represent A as a sequence $(a_k)_{k=1}^{\infty}$. It may happen that reasonable accuracy can be obtained by using one of the n points $(a_k)_{k=1}^{n}$. Improvements in technology allow one to increase n, and achieve greater accuracy and an improved user experience. Improvements in theory may suggest another countable, dense subset $B = \{b_k : k \in \mathbb{N}_+\}$, which permits greater accuracy for a given n than A; greater satisfaction for the same cost.

The prototype of all separable metric spaces is $\mathbb{R}$. The rationals have long been used to calculate with real numbers and this is possible because they form a countable, dense subset. One way to calculate with the rationals as approximations to the reals is to use all rationals numerically less than 10^m, and with denominator less than 10^m. This is the technology of the measuring tape. Another is to use all numbers that can be written using m digits and a floating decimal point. One would use the largest m that current technology can manage. This corresponds roughly to how floating point arithmetic has evolved in response to increasing computing power.

In order to represent curves, pictures and physical objects we will need to model them as elements of a separable metric space. In this context we can expect to see spaces of continuous functions on a compact set, and similar spaces perhaps admitting some discontinuous functions, as natural tools for modelling natural phenomena. So it is important to know which of the classical infinite-dimensional

---

[2] This is written in 2018.

normed spaces are separable, and be familiar with a selection of countable, dense subsets which can provide a palette of approximating objects.

5.2.1   Second Countability

The strangely named topological property has nothing to do with keeping accurate time. It is closely related to separability.

Definition Let (X, d) be a metric space. A family $(U_i)_{i \in I}$ of non-empty open sets is called a base for the topology of X (or a base for the open sets of X) if every open set in X is the union of some subfamily of $(U_i)_{i \in I}$.

It makes little difference whether we think of a base as a family, or as a set $\mathcal{B}$ of open sets, such that every open set of X is a union of some subset of $\mathcal{B}$. An obvious example of a base is the family of all open balls

$$\left(B_r(a) \right)_{r > 0, \, a \in X}.$$

Definition A metric space is said to be *second countable*, or, in lengthier language, it is said to satisfy the *second axiom of countability*, if it has a countable base for its topology.

An example of a second countable space is $\mathbb{R}$. We can use the family, clearly countable, of all intervals $]a, b[$ with rational endpoints as a base.

It turns out that second countability is equivalent to separability for metric spaces (but for topological spaces in general it is more restrictive).

Proposition 5.3 *A metric space is separable if and only if it is second countable.*

Proof Suppose that X is second countable. Let $(U_n)_{n=1}^{\infty}$ be a countable base for the open sets of X. Let $(a_n)_{n=1}^{\infty}$ be a family of points in X, such that $a_n \in U_n$ for each n. Then the set $A = \{a_n : 1 \leq n < \infty\}$ is countable, and dense in X since every open set contains a point of A.

Conversely, suppose that X is separable. Let A be a countable dense subset of X. An obvious candidate for a countable base is the family of all open balls with rational radius and centre in A

$$\mathcal{B} := \left(B_q(a) \right)_{q \in \mathbb{Q}_+, \, a \in A}.$$

In order to show that this is a base it is enough to show that every open ball is the union of balls extracted from this family $\mathcal{B}$. We can infer this if we can show the following: for every ball $B_r(a)$ and every point $b \in B_r(a)$, there exists $c \in A$ and a rational $q > 0$, such that $b \in B_q(c)$ and $B_q(c) \subset B_r(a)$. To achieve this we choose rational q so that

$$0 < 2q < r - d(a, b)$$

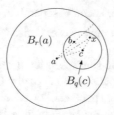

Fig. 5.1 Picture of the proof

and then $c \in A$, such that $d(b, c) < q$ (possible because A is dense). Then $b \in B_q(c)$ and, by the triangle inequality, we have that if $x \in B_q(c)$ then

$$d(x, a) \leq d(x, c) + d(c, b) + d(b, a) < 2q + d(b, a) < r,$$

so that $B_q(c) \subset B_r(a)$. As usual this kind of argument can be illustrated by a picture of Euclidean discs (Fig. 5.1). $\square$

Now we can build a vast number of separable spaces.

Proposition 5.4 *A subspace of a separable space is separable.*

Proof Let X be separable and let M be a subspace of X. At first sight one would like to take the intersection of M with a countable dense subset A of X. But the intersection can be empty (the reader should think of an example), so this doesn't work. The trick is to use proposition 5.3. Let $(U_i)_{i \in I}$ be a base for the open sets of X. Then $(U_i \cap M)_{i \in I}$ is a base for the open sets of M. So if X has a countable base so does M. $\square$

There is also, as the reader no doubt suspects, a first axiom of countability, and spaces satisfying it are called first countable. This asserts that every point has a countable base for its neighbourhoods (see 4.3 exercise 11 for an example involving this concept). More precisely, for each point a, there exists a countable family $(U_n)_{n=1}^{\infty}$ of neighbourhoods of a, such that for any neighbourhood V of a there exists n such that $U_n \subset V$. This is of little interest for us since all metric spaces are first countable.

5.2.2 Exercises

1. Show that metric spaces are first countable.
2. It was claimed, in the proof of proposition 5.4, that the intersection of M with a countable dense subset A of X can be empty. Of course, a single point subspace of X would suffice as an example. Produce a more entertaining example by showing that through a point (a, b) in the plane $\mathbb{R}^2$, with irrational coordinates, there exists a line, every point of which has irrational coordinates. In fact, the set of all such lines is uncountable.

3. Show that a compact metric space is separable.

   Hint. This cropped up in a previous exercise (4.5 exercise 3), but without the term 'separable'.

4. Let X be a separable metric space and let $\mathcal{B} \subset \mathcal{P}(X)$ be pairwise disjoint and contain only open sets. Show that $\mathcal{B}$ is countable.

5. Let X be a separable metric space and let A be a discrete set in X (that is, A is a discrete subspace of X, or, every point in A is isolated in A). Show that A is countable.

6. Let $1 \le p < \infty$. Show that the space ℓ^p is separable.

7. Show that ℓ^∞ is not separable, whereas its subspace c_0 (2.4 exercise 13 and 3.1 exercise 2) is separable.

   Hint. Find an uncountable discrete set in ℓ^∞.

8. a) Show that the Banach space $C_B(\mathbb{R})$ (the space of bounded continuous functions $f : \mathbb{R} \to \mathbb{R}$) is not separable.

   Hint. Use an approach similar to that of the previous exercise.

   b) ($\Diamond$) More generally, if M is a non-compact metric space then the space $C_B(M)$ is not separable.

   Hint. Same hint as for 4.3 exercise 21.

9. Show that the Banach space $B(M)$ (of bounded real functions on a set M) is separable if and only if M is finite.

10. Let X and Y be metric spaces and suppose that X is separable. Let $f : X \to Y$ be a mapping with the property that $\lim_{x \to a} f(x)$ exists for all $a \in X$ that are limit points of X. Show that the discontinuities of f are at most countably many.

   Hint. Recall the point oscillation ω_f of the function f (section 2.3.3). Using proposition 2.27 show that for all $\varepsilon > 0$, the set $\omega_f^{-1}(]\varepsilon, \infty[)$ is discrete. Apply proposition 2.26 and the result of exercise 5.

11. Let X and Y be separable spaces. Prove that $X \times Y$ is separable.

12. Show that a family of open sets $(U_i)_{i \in I}$ is a base for the open sets of a metric space X if and only if the following condition is satisfied: for every open set A and every $x \in A$ there exists i such that $x \in U_i \subset A$.

13. Let $(X_n)_{n=1}^\infty$ be a sequence of metric spaces. The object is to describe a base for the open sets of the product $\prod_{n=1}^\infty X_n$ and deduce that the latter is separable if each factor X_n is separable.

   For each n, let $\mathcal{A}_n$ be a base for the open sets of X_n. Let $\mathcal{B}$ be the set of all subsets of $\prod_{n=1}^\infty X_n$ of the form $\prod_{n=1}^\infty U_n$, where

   i) There exists N such that $U_n \in \mathcal{A}_n$ for $1 \le n \le N$.
   ii) $U_n = X_n$ for all $n > N$.

   Show that:

   a) $\mathcal{B}$ is a base for the open sets of $\prod_{n=1}^\infty X_n$.
   b) If each $\mathcal{A}_n$ is countable then so is $\mathcal{B}$.
   c) If each X_n is separable then so is $\prod_{n=1}^\infty X_n$.

14. Construct explicitly a countable dense subset of $\mathbb{R}^{\mathbb{N}+}$.

    Note. It's not $\mathbb{Q}^{\mathbb{N}+}$. That's not countable.

15. In the presence of separability, conclusions that in their full generality depend on the axiom of choice, can often be obtained without using it. By a choice function for a set $\mathcal{B}$ whose elements are themselves sets, we mean a function $g : \mathcal{B} \to \bigcup \mathcal{B}$, such that $g(B) \in B$ for each $B \in \mathcal{B}$. The axiom of choice guarantees that a choice function exists if the elements of $\mathcal{B}$ are non-empty sets. In the following cases $\mathcal{B} \subset \mathcal{P}(M)$ where M is a metric space. Define a choice function without using the axiom of choice:

    a) If M is a separable metric space and the elements of $\mathcal{B}$ are open non-empty sets.
    b) If $M = \mathbb{R}^n$ and the elements of $\mathcal{B}$ are compact non-empty sets.
    c) If $M = \mathbb{R}^n$ and the elements of $\mathcal{B}$ are closed non-empty sets.

    Does item (a) throw any light on the purported need for the axiom of choice in the proof of proposition 4.1?

16. Let X be a separable metric space. Show that every open covering has a countable subcovering. More precisely, if

$$X = \bigcup_{i \in I} U_i,$$

    where the sets U_i are all open, then there exists a countable subset $J \subset I$, such that

$$X = \bigcup_{i \in J} U_i.$$

    Hint. Let $(B_n)_{n=1}^{\infty}$ be a countable base for the topology of X (proposition 5.3). For each $x \in X$ there exists $i \in I$ and $n(x) \in \mathbb{N}_+$, such that $x \in B_{n(x)} \subset U_i$. The set $\{n(x) : x \in X\}$ is countable. Use it to select a countable subfamily of the family $(U_i)_{i \in I}$ that covers X.

    Note. The result is called Lindelöf's theorem. A topological space with the property, obviously topological, that every open covering has a countable subcovering, is called a Lindelöf space.

17. Let X be a separable metric space. The object is to build a homeomorphism of X on to a subspace of the Hilbert cube H^{∞}. Replace the metric d of X by an equivalent metric taking values in the interval $[0, 1]$, for example $d'(x, y) := \min(d(x, y), 1)$. Now let $(a_n)_{n=1}^{\infty}$ be a dense sequence in X. Consider the mapping

$$F : X \to H^{\infty}, \quad F(x) = (d'(x, a_n))_{n=1}^{\infty}.$$

Prove that F is a homeomorphism of X on to a subspace of the Hilbert cube H^∞.

Hint. Show in succession:

a) F is injective;
b) F is continuous; and
c) $F^{-1} : F(X) \to X$ is continuous.

One way to tackle continuity is to use the criterion by sequences, proposition 2.29, and pay no attention to the metric in H^∞.

Note. In particular (by exercise 3) a compact metric space is homeomorphic to a subspace of H^∞. In case X is compact the continuity of the inverse of F follows automatically by proposition 4.10.

18. Show that a separable metric space has cardinality at most **c**.

Hint. Use exercise 17. What is the cardinality of H^∞? You may need some simple cardinal arithmetic.

5.3 ($\Diamond$) Weierstrass

In this section we make a sizeable excursion into analysis to describe one of the numerous proofs that an important classical space is separable, and provide an example of a countable dense subset that has practical importance.

Proposition 5.5 (Weierstrass Approximation Theorem) *Let* $f \in C[a, b]$ *and* $\varepsilon > 0$. *Then there exists a polynomial* $p(x)$, *such that*

$$\sup_{a \le x \le b} |f(x) - p(x)| < \varepsilon.$$

Let us note at once the following consequences, which have both theoretical and practical value:

a) The space of all polynomials is dense in $C[a, b]$. For the inequality says that $\|f - p\|_\infty < \varepsilon$.
b) The space $C[a, b]$ is separable. This is because every polynomial can be uniformly approximated on $[a, b]$ by another polynomial that has rational coefficients. The space of all polynomials with rational coefficients is countable.
c) Let f be continuous on $[a, b]$ and real valued. Then there exists a sequence of polynomials $(p_n)_{n=1}^\infty$ such that $p_n \to f$ uniformly on the interval $[a, b]$.

The rest of this section is devoted to a proof of the Weierstrass approximation theorem. The proof will also provide a practical method for approximating a given continuous function by a polynomial.

The first thing to note is that if we can prove the theorem for the interval $[0, 1]$ then it can be deduced for the interval $[a, b]$. Suppose that we know how to approximate continuous functions on $[0, 1]$ uniformly by polynomials. Now given a continuous function f on $[a, b]$ we can approximate the function

$$g(x) := f(x(b - a) + a), \quad (0 \le x \le 1)$$

uniformly by polynomials on $[0, 1]$. Suppose, for example, that $p_n(x) \to g(x)$ uniformly for $0 \le x \le 1$. Then

$$p_n\left(\frac{x - a}{b - a}\right) \to f(x)$$

uniformly for $a \le x \le b$. It therefore suffices to prove the theorem for the interval $[0, 1]$.

Bernstein Polynomials

A Bernstein[3] polynomial is by definition a linear combination of polynomials of the form $x^k(1 - x)^{n-k}$. In the following calculations we make much use of the binomial coefficients $\binom{n}{k}$ and the binomial theorem, so the reader is advised to recall them.

Definition Let $f \in C[0, 1]$. The Bernstein polynomial of f of degree n is the polynomial

$$(B_n f)(x) = \sum_{k=0}^{n} f\left(\frac{k}{n}\right)\binom{n}{k}x^k(1 - x)^{n-k}.$$

Proof of Proposition 5.5 We are going to prove the Weierstrass approximation theorem by showing that $B_n f \to f$ uniformly on $[0, 1]$. We begin with three special cases of f. Let f_0, f_1 and f_2 be the functions $1, x$ and x^2.

Step 1. The case $f = f_0$. We have, using the binomial theorem:

$$(B_n f_0)(x) = \sum_{k=0}^{n} \binom{n}{k}x^k(1 - x)^{n-k} = (x + (1 - x))^n = 1 = f_0(x).$$

Step 2. The case $f = f_1$. We have:

$$(B_n f_1)(x) = \sum_{k=0}^{n} \frac{k}{n}\binom{n}{k}x^k(1 - x)^{n-k}$$

$$= \sum_{k=1}^{n} \binom{n - 1}{k - 1}x^k(1 - x)^{n-k}$$

---

[3] S. N. Bernstein, not F. Bernstein of the Schröder-Bernstein theorem.

$$= \sum_{k=0}^{n-1} \binom{n-1}{k} x^{k+1} (1-x)^{n-k-1}$$

$$= x \sum_{k=0}^{n-1} \binom{n-1}{k} x^{k} (1-x)^{n-k-1}$$

$$= x \big(x + (1-x) \big)^{n-1} \quad \text{(by the binomial theorem)}$$

$$= x.$$

Step 3. The case $f = f_2$. For $k = 0, 1, \ldots, n$ we have (think of $\binom{n}{j}$ as 0 if $j < 0$):

$$\left(\frac{k}{n}\right)^2 \binom{n}{k} = \frac{k}{n} \binom{n-1}{k-1}$$

$$= \left(\left(1 - \frac{1}{n}\right)\left(\frac{k-1}{n-1}\right) + \frac{1}{n}\right) \binom{n-1}{k-1}$$

$$= \left(1 - \frac{1}{n}\right) \binom{n-2}{k-2} + \frac{1}{n} \binom{n-1}{k-1}$$

We obtain, again using the binomial theorem:

$$(B_n f_2)(x) = \sum_{k=0}^{n} \left(\frac{k}{n}\right)^2 \binom{n}{k} x^{k} (1-x)^{n-k}$$

$$= \left(1 - \frac{1}{n}\right) \sum_{k=2}^{n} \binom{n-2}{k-2} x^{k} (1-x)^{n-k}$$

$$+ \frac{1}{n} \sum_{k=1}^{n} \binom{n-1}{k-1} x^{k} (1-x)^{n-k}$$

$$= \left(1 - \frac{1}{n}\right) x^2 + \frac{1}{n} x$$

Hence

$$B_n f_2 - f_2 = \frac{1}{n} (x - x^2)$$

and so $B_n f_2 - f_2 \to 0$ uniformly for $0 \le x \le 1$.

Step 4. We need the following lemma:

Lemma 5.1 *Let $\delta > 0$, let $x \in [0, 1]$ and let n be a positive integer. Let F be the set of all $k \in \{0, 1, \ldots, n\}$, such that $\left|\frac{k}{n} - x\right| \geq \delta$. Then*

$$\sum_{k \in F} \binom{n}{k} x^k (1 - x)^{n-k} \leq \frac{1}{4n\delta^2}$$

Proof Firstly

$$\sum_{k \in F} \binom{n}{k} x^k (1 - x)^{n-k} \leq \frac{1}{\delta^2} \sum_{k \in F} \left(\frac{k}{n} - x\right)^2 \binom{n}{k} x^k (1 - x)^{n-k}$$

$$\leq \frac{1}{\delta^2} \sum_{k=0}^{n} \left(\frac{k}{n} - x\right)^2 \binom{n}{k} x^k (1 - x)^{n-k}.$$

Secondly, we transform the right-hand member

$$\frac{1}{\delta^2} \sum_{k=0}^{n} \left(\frac{k}{n} - x\right)^2 \binom{n}{k} x^k (1 - x)^{n-k}$$

$$= \frac{1}{\delta^2} \sum_{k=0}^{n} \left(\left(\frac{k}{n}\right)^2 - 2\frac{k}{n}x + x^2\right) \binom{n}{k} x^k (1 - x)^{n-k}$$

$$= \frac{1}{\delta^2} \left((B_n f_2)(x) - 2x(B_n f_1)(x) + x^2(B_n f_0)(x)\right).$$

$$= \frac{1}{\delta^2} \left(\left(1 - \frac{1}{n}\right)x^2 + \frac{1}{n}x - 2x^2 + x^2\right)$$

$$= \frac{1}{n\delta^2} x(1 - x) \leq \frac{1}{4n\delta^2}.$$

This proves the lemma. □

Step 5 and conclusion of the proof of proposition 5.5.
Let $f \in C[0, 1]$. We can write (again using the binomial theorem):

$$f(x) - (B_n f)(x) = f(x) - \sum_{k=0}^{n} f\left(\frac{k}{n}\right) \binom{n}{k} x^k (1 - x)^{n-k}$$

$$= \sum_{k=0}^{n} \left(f(x) - f\left(\frac{k}{n}\right)\right) \binom{n}{k} x^k (1 - x)^{n-k}.$$

We have to show that the right-hand side converges to 0 uniformly on $[0, 1]$ as $n \to \infty$.

Let $\varepsilon > 0$. Now f is uniformly continuous on $[0, 1]$. Hence there exists $\delta > 0$, such that $|f(x) - f(y)| < \varepsilon$ for all x and y in $[0, 1]$ that satisfy $|x - y| < \delta$. For a given n and $x \in [0, 1]$ we let $F \subset \{0, 1, \ldots, n\}$ be the set of those k, such that $\left| x - \frac{k}{n} \right| \geq \delta$. Then

$$\left| f(x) - (B_n f)(x) \right| \leq \sum_{k=0}^{n} \left| f(x) - f\left(\frac{k}{n}\right) \right| \binom{n}{k} x^k (1 - x)^{n-k}$$

$$\leq \varepsilon \sum_{k \notin F} \binom{n}{k} x^k (1 - x)^{n-k}$$

$$+ \sum_{k \in F} \left| f(x) - f\left(\frac{k}{n}\right) \right| \binom{n}{k} x^k (1 - x)^{n-k}$$

$$\leq \varepsilon \sum_{k=0}^{n} \binom{n}{k} x^k (1 - x)^{n-k} + 2\|f\|_\infty \sum_{k \in F} \binom{n}{k} x^k (1 - x)^{n-k}$$

$$\leq \varepsilon + \frac{\|f\|_\infty}{2n\delta^2} \quad \text{(by the lemma)}$$

Now δ depends only on ε and f (just look back to see where δ was introduced). If $n \geq \|f\|_\infty / 2\delta^2 \varepsilon$ we find $\|f - B_n f\|_\infty < 2\varepsilon$. We conclude that $B_n f \to f$ uniformly on $[0, 1]$. $\square$

This proof of the Weierstrass approximation theorem is closely connected to probability theory. The reader may know that the quantity $\binom{n}{k} x^k (1 - x)^{n-k}$ is the probability of k successes in n Bernoulli trials, when the probability of success in one trial is x. The inequality in the lemma is Chebyshev's inequality.

5.3.1 Exercises

1. Show that every polynomial can be approximated uniformly on the interval $[a, b]$ by a polynomial with rational coefficients. Alongside the Weierstrass approximation theorem this shows that the space $C[a, b]$ is separable.
2. Show that $B_n(f)$ (as defined in the proof of proposition 5.5) lies beween the maximum and minimum of f in the interval $[0, 1]$.
3. Show that the space $C(\mathbb{R})$ (see 3.3 for the definition) is separable.

   Note. Compare 5.2 exercise 8(a). The "smaller" space $C_B(\mathbb{R})$ is not separable. Of course, the metrics are quite different. See the next exercise.
4. Show that the injection $J : C_B(\mathbb{R}) \to C(\mathbb{R})$, $J(f) = f$, is continuous, but its inverse defined on its range, is discontinuous.

   Note. Putting it differently, every set open in $J(C_B(\mathbb{R}))$ is open in $C_B(\mathbb{R})$, but there are sets open in $C_B(\mathbb{R})$ that are not open in $J(C_B(\mathbb{R}))$. Having more open sets makes it harder to be separable. One says that the topology of $C_B(\mathbb{R})$ is *finer* than the topology of $J(C_B(\mathbb{R}))$.

5. Show that the set $C(\mathbb{R})$ has cardinality **c**.

   Hint. Use exercise 3, 5.2 exercise 18 and the Schröder-Bernstein theorem of set theory.

6. Show that the Banach space $C_0(\mathbb{R})$ (of continuous functions that vanish at infinity, 2.2 exercise 19 and 3.3 exercise 8) is separable.

   Hint. Transform the line $\mathbb{R}$ into the interval $]-1, 1[$, for example by using the hyperbolic tangent function.

7. Let $f : [0, 1] \to \mathbb{R}$ be continuous and suppose that $\int_0^1 f(x)x^n \, dx = 0$ for every natural number n. Show that $f = 0$.

   Hint. Show that $\int_0^1 f(x)p(x) \, dx = 0$ for all polynomials $p(x)$ and approximate f uniformly by a polynomial.

8. Let $(x_n)_{n=1}^\infty$ be a sequence of points in the interval $[0, 1]$ with the property that

$$\lim_{n \to \infty} \frac{1}{n} \sum_{k=1}^n x_k^p$$

exists for all natural numbers p. Show that

$$\lim_{n \to \infty} \frac{1}{n} \sum_{k=1}^n f(x_k)$$

exists for every $f \in C[0, 1]$.

   Hint. Use proposition 5.1.

9. Bernstein polynomials have turned out to be important in the area of computer graphics. We refer here to the so-called Bézier curves. Many of the graphics in this text were drawn by a hand-held device (a "mouse") that controls a computer application that creates such curves.

   For integers $n \geq 0$ and k, such that $0 \leq k \leq n$, we set

$$B_{k,n}(t) = \binom{n}{k} t^k (1 - t)^{n-k}$$

   Now let $P_0, P_1, \ldots, P_n$ be distinct points in a coordinate plane $\mathbb{R}^2$. The Bézier curve determined by the points P_k is the parametric curve

$$X(t) = \sum_{k=0}^n B_{k,n}(t) P_k.$$

a) Show that the Bézier curve determined by the points P_0, P_1, ..., P_n lies in their convex hull (2.2 exercise 20).

b) Show that the tangent to the curve at the initial point P_0 is the line through P_0 and P_1, whilst the tangent line through the final point P_n is the line through P_{n-1} and P_n.

5.3.2   Pointers to Further Study

$\rightarrow$ Stone-Weierstrass theorem
$\rightarrow$ Numerical analysis
$\rightarrow$ Probability theory
$\rightarrow$ Computer graphics

5.4   ($\Diamond$) Stone-Weierstrass

We study a remarkable extension of the Weierstrass approximation theorem, in which conditions are given that guarantee that a certain kind of subset A of the space $C(M)$ is dense in $C(M)$. Here M is supposed to be a compact metric space and, as usual, $C(M)$ is the space of all continuous real functions on M, with the norm

$$\|f\| = \sup_M |f|.$$

The proof is a nice example of what can be done using the open-covering definition of compactness (as opposed to sequential compactness). Although the Stone-Weierstrass theorem includes the Weierstrass approximation theorem as a special case, the proof could hardly be more different. Whereas the proof, given in the last section, of the Weierstrass approximation theorem was completely constructive, offering a practical method for building the approximations, the proof we present of the Stone-Weierstrass theorem is non-constructive, probably of necessity, owing to its great generality. This difference of character fully justifies the inclusion in this text of both theorems.

We will need one preliminary fact drawn from the classical Weierstrass approximation theorem, that there exists a sequence of polynomials $(u_n(t))_{n=1}^\infty$, that converges to the function $v(t) := |t|$ uniformly on the interval $[-1, 1]$.

Actually one might try to prove this preliminary fact without appealing to the Weierstrass approximation theorem. This would make the ensuing proof also a proof of the Weierstrass theorem. One way is to show that the Maclaurin series of the function $(1-t)^{1/2}$ converges uniformly on the interval $[0, 1]$. The series is of course the binomial series and certainly converges to $(1-t)^{1/2}$ in $[0, 1[$; one would have to show that it is also convergent at $t = 1$, and the convergence is uniform on the closed interval, something that essentially follows from Abel's theorem on power series. We end up with a sequence of polynomials $v_n(t)$ (partial sums of the Maclaurin series) that converge to $(1 - t)^{1/2}$ uniformly with respect to t in $[0, 1]$, and then the polynomials $u_n(t) := v_n(1 - t^2)$ converge to $|t|$ uniformly on $[-1, 1]$.

Recall that a subset $A \subset C(M)$ is an *algebra* if it is a linear subspace, and for all f and g in A the product fg is in A.

A set $A \subset C(M)$ is said to *separate the points of* M, if, for all s and t in M there exists $f \in A$, such that $f(s) \neq f(t)$.

In the ensuing pages, take care not to confuse $\| f \|$, that is, the norm of f in $C(M)$, and $|f|$, that is, the function $x \mapsto |f(x)|$, an element of $C(M)$.

Proposition 5.6 (Stone-Weierstrass Theorem) *Let M be compact and let $A \subset C(M)$ be an algebra that contains the function 1 and separates the points of M. Then A is dense in $C(M)$.*

Proof The proof is in several steps. We first set out some preliminary observations.

We have to show that $\overline{A} = C(M)$. Now it is easy to show that $\overline{A}$ is also an algebra (it is left to the reader to check this point). Hence we may as well assume that A is in fact closed from the start, the object now being to show that $A = C(M)$. So from now on we assume that A is closed in $C(M)$.

Since 1 is in A it follows that A includes all constants. Moreover, as the reader should check, if $h(t)$ is a polynomial and $f \in A$, then $h \circ f \in A$.

Step 1. We prove that if $f \in A$, then $|f| \in A$.

Let u_n ($n = 1, 2, 3, \ldots$) be polynomials, such that $u_n(t) \to |t|$ uniformly on the interval $[-1, 1]$. Let $a = \| f \|$. If $a = 0$ the required conclusion is obvious.

Assume then that $a > 0$. The function f/a only takes values in the interval $[-1, 1]$, and therefore the functions $u_n \circ (f/a)$, which belong to A, converge to $|f/a| = |f|/a$ uniformly on M. We conclude, since we are assuming that A is closed (see the preliminary observations), that $|f| \in A$.

Step 2. We show that if f and g are in A, then the functions $\min(f, g)$ and $\max(f, g)$ also belong to A.

We have that $\min(f, g) = \frac{1}{2}(f + g - |f - g|)$ and $\max(f, g) = \frac{1}{2}(f + g + |f - g|)$, so that this follows from step 1.

Step 3. Let s and t be distinct points of M, and let α and β be real numbers. We show that there exists $f \in A$, such that $f(s) = \alpha$ and $f(t) = \beta$.

By assumption there exists $g \in A$, such that $g(s) \neq g(t)$. Now it is easy to find λ and μ, such that $\lambda g(s) + \mu = \alpha$ and $\lambda g(t) + \mu = \beta$. The function $\lambda g + \mu$ belongs to A.

Step 4. Let $f \in C(M)$, $x \in M$ and $\varepsilon > 0$. We show that there exists $h \in A$, such that $h(x) = f(x)$, whilst for all $y \in M$ we have $h(y) < f(y) + \varepsilon$.

According to step 3, for each $t \in M$ we can find a function $g_t \in A$, such that $g_t(x) = f(x)$ and $g_t(t) = f(t)$. The set

$$U_t := \{y \in M : g_t(y) < f(y) + \varepsilon\}$$

is open in M and contains t. Since M is compact there exists a finite set $\{t_1, \ldots, t_m\}$, such that the sets $U_{t_1}, \ldots, U_{t_m}$ cover M. Finally we put

$$h = \min(g_{t_1}, \ldots, g_{t_m}).$$

By step 2 the function h is in A. Also $h(x) = f(x)$. For each y there exists k, such that $y \in U_{t_k}$, and therefore $h(y) \leq g_{t_k}(y) < f(y) + \varepsilon$.

Step 5. End of the proof. Let $f \in C(M)$. We show that $f \in A$ by showing that $f \in \overline{A}$.

Let $\varepsilon > 0$. By step 4, for each $x \in M$ there exists $h_x \in A$, such that $h_x(x) = f(x)$, whilst for each y we have $h_x(y) < f(y) + \varepsilon$. The set

$$U_x = \{t \in M : h_x(t) > f(t) - \varepsilon\}$$

is open in M and contains x. Since M is compact there exists a finite set $\{x_1, \ldots, x_m\}$, such that the sets $U_{x_1}, \ldots, U_{x_m}$ cover M. Put

$$g = \max(h_{x_1}, \ldots, h_{x_m}).$$

Now g is in A by step 2, and we can show that it satisfies

$$f(y) - \varepsilon < g(y) < f(y) + \varepsilon$$

for all $y \in M$. In fact, given $y \in M$ there exists k, such that $y \in U_{x_k}$. Then

$$g(y) \geq h_{x_k}(y) > f(y) - \varepsilon$$

and since $h_{x_k}(y) < f(y) + \varepsilon$ for each k, we also have

$$g(y) < f(y) + \varepsilon.$$

This shows that we can approximate f uniformly by elements of A to arbitrary precision. That is, $f \in \overline{A}$, and since A is closed we have $f \in A$. $\square$

5.4.1 Exercises

1. Let M be a closed and bounded subset of $\mathbb{R}^n$. Show that the space of polynomials in n variables is dense in $C(M)$.
2. Let M be the product $X \times Y$, where X and Y are compact metric spaces. Show that every continuous function $f : M \to \mathbb{R}$ can be approximated uniformly with arbitrary accuracy by a function of the form

$$h(x, y) := \sum_{k=1}^{m} u_k(x) v_k(y)$$

for an integer m, and functions $u_k \in C(X)$ and $v_k \in C(Y)$, $k = 1, \ldots, m$.
3. The complex-valued continuous functions on a compact space M form a vector space $C_{\mathbb{C}}(M)$ over $\mathbb{C}$ and an algebra. This is metrised by setting $d(f, g) = \sup_M |f - g|$ and interpreting the absolute value as the modulus of a complex number (or, what is the same, Euclidean distance in $\mathbb{R}^2$). A subalgebra A of $C_{\mathbb{C}}(M)$ that contains the constants and separates points may fail to be dense in $C_{\mathbb{C}}(M)$. Nevertheless, a further small condition will suffice for A to be dense.

   a) For this item you will need some knowledge of complex analysis. Let M be the closed unit disc $\{z \in \mathbb{C} : |z| \leq 1\}$. The algebra consisting of all polynomials $f(z)$ in the variable z, with complex coefficients, separates points in M and contains all constants. Yet it is not dense in $C_{\mathbb{C}}(M)$. Show, for example, that its closure does not contain the function $\bar{z}$.

   b) Let M be a compact metric space. Let A be a subalgebra of $C_{\mathbb{C}}(M)$ that contains all constants, separates points and, in addition, has the following property: for each $f \in A$ the complex conjugate $\bar{f}$ is in A. Prove that A is dense in $C_{\mathbb{C}}(M)$.

   Hint. Show that the spaces $\mathrm{Re}\,(A)$ and $\mathrm{Im}\,(A)$ are dense in the real space $C_{\mathbb{R}}(M)$.

4. a) Show that the algebra of all polynomials $f(z, \bar{z})$, in the variables z and $\bar{z}$, is dense in $C_{\mathbb{C}}(T)$, where T is the unit circle $\{z \in \mathbb{C} : |z| = 1\}$.

   Hint. Use the previous exercise.

   b) Deduce from item (a) that the space of all real trigonometric polynomials, that is, the set of all functions of the form

$$a_0 + \sum_{k=1}^{m} (a_k \cos kx + b_k \sin kx)$$

   with real coefficients, is dense in the space of all continuous, real 2π-periodic functions of x.

Note. The space of continuous, real 2π-periodic functions carries the supremum norm. It can be identified with the subspace of $C[0, 2\pi]$ consisting of functions satisfying $f(0) = f(2\pi)$.

c) Let f be continuous and 2π-periodic, and suppose that

$$\int_0^{2\pi} f(x) \cos nx \, dx = \int_0^{2\pi} f(x) \sin nx \, dx = 0$$

for all natural numbers n. Show that $f = 0$.

Hint. The assumptions imply that $\int_0^{2\pi} f(x) T(x) \, dx = 0$ for every trigonometric polynomial $T(x)$.

Note. Items (b) and (c) are key facts underlying the theory of Fourier series.

5. Let M be a compact metric space. We know that there exists a countable dense subset D in M (see 5.2 exercise 3). Enumerate D as a sequence $(a_k)_{k=1}^{\infty}$. Define $f_k \in C(M)$ by $f_k(x) = d(x, a_k)$. Show that the algebra spanned by the functions $f_k, (k = 1, 2, \ldots)$ is dense in $C(M)$. Deduce from this that $C(M)$ is separable.

6. Let M be a compact non-empty metric space. Show that the cardinality of $C(M)$ equals **c**.

5.4.2 *Pointers to Further Study*

$\rightarrow$ Korovkin's theorem
$\rightarrow$ Fourier series

Chapter 6
Properties of Complete Spaces

Existence is not a predicate

E. Kant. Critique of Pure Reason

Compactness and completeness stand side by side as the two most important properties that a metric space can possess. Both have a striking role in analysis in proving that desirable objects exist. These range from solutions to equations (including differential equations, both ordinary and partial) and solutions to extremal problems, to equilibrium strategies in game theory.

The importance of completeness is such, that an incomplete space may be considered defective, just as $\mathbb{Q}$ is defective because there are no irrationals. The Lebesgue integral replaced the Riemann integral partly because using it one can build complete normed spaces of integrable functions. This facilitated the theory of distributions and the complete function spaces (Sobolev spaces) of the theory of partial differential equations.

6.1 Cantor's Nested Intersection Theorem

The following proposition appeared originally as a basic property of the real numbers, and its conclusion in that context has sometimes been used as an axiom to express the Dedekind completeness of the real number field (although it must then be supplemented by the Archimedean axiom, asserting that the natural numbers are not bounded above).

Proposition 6.1 *Let (X, d) be a complete metric space. Let $(A_n)_{n=1}^{\infty}$ be a sequence of subsets of X that satisfy the following conditions:*

1) *Each set A_n is closed and non-empty.*
2) $A_{n+1} \subset A_n$ *for each n.*
3) $\operatorname{diam}(A_n) \to 0.$

Then the intersection $\bigcap_{n=1}^{\infty} A_n$ is non-empty and consists of exactly one point.

© The Author(s), under exclusive license to Springer Nature Switzerland AG 2022   187
R. Magnus, *Metric Spaces*, Springer Undergraduate Mathematics Series,
https://doi.org/10.1007/978-3-030-94946-4_6

Proof Choose a point x_n in A_n for each n. If $m > n$ we have $d(x_m, x_n) \leq$ diam (A_n), and by condition 3 we see that $(x_n)_{n=1}^{\infty}$ is a Cauchy sequence. Hence the sequence $(x_n)_{n=1}^{\infty}$ is convergent. Let $y = \lim_{n \to \infty} x_n$. By condition 2, for each m the sequence $(x_n)_{n=1}^{\infty}$ is eventually in A_m, which is closed by condition 1. Hence $y \in A_m$. We therefore have $y \in \bigcap_{n=1}^{\infty} A_n$. Finally if $y' \in \bigcap_{n=1}^{\infty} A_n$ then $d(y, y') \leq$ diam (A_n) for each n, and by condition 3 we find that $d(y, y') = 0$. That is, $y = y'$.   □

Notes About Cantor's Theorem

Some of the claims, made in the following notes, are explored in the exercises.

a) The conclusion, that $\bigcap_{n=1}^{\infty} A_n$ is not empty, does not necessarily hold if condition 3 is dropped. The reader should find a counterexample, where, in the absence of condition 3 the intersection is empty.

b) If, on the other hand, the sets A_n are compact, then condition 3 may be dropped, and the conclusion that $\bigcap_{n=1}^{\infty} A_n$ is not empty will still stand. However the intersection may contain more than one point. We don't even need X to be complete to draw these conclusions.

c) The property of the metric space X expressed by this proposition is equivalent to completeness. More precisely, if a space X has the *nested intersection property*, that $\bigcap_{n=1}^{\infty} A_n \neq \emptyset$ if the sets A_n satisfy the conditions 1, 2 and 3, then X is complete. See the exercises.

d) Cantor used the nested intersection property of the real number field to give one of several proofs that $\mathbb{R}$ is uncountable. See the exercises.

6.1.1   Categories

Let (X, d) be a metric space. We study here a number of properties that have been found very useful for the more advanced theory of Banach spaces.

Definitions

i) A subset $A \subset X$ is said to be *nowhere dense* in X if the interior of $\overline{A}$ is empty.

ii) A subset $A \subset X$ is said to be *of the first category* in X if it is the union of a countable family of nowhere dense sets.

iii) A subset $A \subset X$ is said to be *of the second category* in X if it is not of the first category.

The properties defined here are properties of sets in a metric space, like being open or closed. They are not properties of metric spaces. This point should be kept clear. One does not say "the metric space X is nowhere dense" or "the metric space X is of the second category."

It is hard to see what being nowhere dense actually entails. It is therefore helpful to rewrite the notion, as we do next.

Proposition 6.2 *Let X be a metric space and let $A \subset X$. Then A is nowhere dense if and only the complement of $\overline{A}$ is dense in X.*

Proof We have the following two-way deductive chain: $(\overline{A})^c$ is dense in X $\Leftrightarrow$ every non-empty open set meets $(\overline{A})^c$ $\Leftrightarrow$ no non-empty open set is a subset of $\overline{A}$ $\Leftrightarrow$ A is nowhere dense. $\qquad\qquad\square$

We can paraphrase the proposition as follows: arbitrarily near to any given point is an interior point of the complement of A. It seems to assert that a nowhere dense set is easily avoided and conjures up a nice picture. Wherever we land in X we can move an arbitrarily small distance and find a peaceful space wholly outside A.

Yet another version of the same picture is this: every non-empty open set U includes a non-empty open set V, such that $A \cap V = \emptyset$. Hence we can find an open ball $B_r(a)$ such that $B_r(a) \subset U$ and $B_r(a) \cap A = \emptyset$. Reducing r if necessary we can even ensure that $\overline{B_r(a)} \subset U$ and $\overline{B_r(a)} \cap A = \emptyset$. In this form it can work in tandem with Cantor's nested intersection theorem, provided completeness is to hand.

Proposition 6.3 *Let (X, d) be a complete metric space. Let W be a set of the first category in X. Then the complement of W is dense in X.*

Proof It is enough to show that every non-empty open set contains a point of W^c. Let U be a non-empty open set. Let $W = \bigcup_{n=1}^{\infty} A_n$ where each set A_n is nowhere dense. There exists an open ball B_1, such that $\overline{B_1} \subset U$ and $\overline{B_1} \cap A_1 = \emptyset$. We may furthermore suppose that diam $B_1 < 1$. We now define inductively a sequence of open balls B_n, $n = 1, 2, 3, \ldots$, such that for each n we have

a) $\overline{B_{n+1}} \subset B_n$.
b) diam $B_n < 1/n$.
c) $\overline{B_n} \cap A_n = \emptyset$.

By Cantor's nested intersection theorem, the intersection $\bigcap_{n=1}^{\infty} \overline{B_n}$ consists of exactly one point y, and by construction $y \in U$ and $y \notin \bigcup_{n=1}^{\infty} A_n$; that is $y \in W^c$. $\square$

A dense set cannot be empty; that's a truism. But that is the bit of proposition 6.3 that often gets recorded in the following form:

If (X, d) is a complete metric space then the set X is of second category.

This is often quoted in the following way, which should be disparaged as misleading:

A complete metric space is of second category in itself.

The strange phrase 'in itself' is probably used to avoid the completely mistaken:

A complete metric space is of second category.

This would be wrong because being of second category is not a property of metric spaces; it is a property of sets in metric spaces, like being open or closed. However we can define a property of metric spaces, plainly a topological property:

Definition A metric space (X, d) is called a Baire space if the set X is of second category.

Proposition 6.3 implies that all complete metric spaces are Baire spaces (indeed the proposition is often called the Baire category theorem). Any space homeomorphic to a complete metric space is also a Baire space; some of them are even wildly incomplete. For example the space of irrational numbers, a subspace of $\mathbb{R}$, is a Baire space, according to 3.2 exercise 9.

Thoughts About the Proof

The axiom of choice is needed to clarify precisely the inductive definition of the sequence of open balls B_n in the proof of proposition 6.3. The reader may feel that the proof as given, so simple and decisive, needs no further clarification, and that we are embarking on obfuscation. If so, they may omit the following two paragraphs. The situation is similar to, but simpler than, the case of proposition 4.13; however, there is an interesting twist in the last paragraph. The context and notation are as in the statement and proof of proposition 6.3.

To each pair (n, V), where n is a natural number and V a non-empty open set, we assign the set $\mathcal{S}_{(n,V)}$ of all open balls B, with diameter δ satisfying $\delta < 1/n$, and such that $\overline{B} \subset V$ and $\overline{B} \cap A_n = \emptyset$. The set $\mathcal{S}_{(n,V)}$ is non-empty by proposition 6.2. Note that when we say that B is an open ball, we mean that it has a representation of the form $B = B_r(x)$, for some $r > 0$ and $x \in X$ (and is thus not empty). The set B does not necessarily determine r and x uniquely, a problem we avoid by concentrating on B as a set, and measuring it by its diameter (which is well defined but may be 0). Now we need a choice function κ, that assigns to each non-empty set of open balls one of its members. The inductive step in the proof entails setting

$$B_{n+1} = \kappa\left(\mathcal{S}_{(n+1, B_n)}\right).$$

In spite of this, the reader might consult proposition 5.3, and conclude that if X is *separable*, then we can dispense with the axiom of choice in the proof of proposition 6.3, since we can restrict ourselves to working only with open balls selected from a countable base. It is a simple matter, left to the reader, to show that for subsets of a countable set, a choice function can be constructed without the axiom of choice.

6.1.2 Exercises

1. Verify notes (a) and (b) (in 'Notes about Cantor's theorem').
2. Suppose that a metric space X has the nested intersection property. Show that X is complete. (This is note (c)).

Hint. Suppose that $(a_n)_{n=1}^{\infty}$ is a Cauchy sequence. Consider the sets

$$A_n = \overline{\{a_k : k \geq n\}}, \quad n = 1, 2, \ldots$$

3. Suppose that $(a_k)_{k=1}^{\infty}$ is a sequence in the interval $[0, 1]$. Of the three intervals $[0, \frac{1}{3}]$, $[\frac{1}{3}, \frac{2}{3}]$, $[\frac{2}{3}, 1]$, choose one that does not contain a_1. Now divide the chosen interval into three of the same length and choose one that does not contain a_2. Proceed inductively to define a decreasing chain of closed intervals whose lengths tend to zero, and are such that the nth interval does not contain a_n. Use this to produce an element of $[0, 1]$ that is not in the sequence $(a_k)_{k=1}^{\infty}$. This was one of Cantor's proofs of the uncountability of the real numbers (and was note (d)).

4. That 'first category' and 'second category' are not predicates applicable to metric spaces is clarified by a simple example, demonstrating that there is no sense in talking about the category of the metric space $\mathbb{N}$. What is the category of $\mathbb{N}$ when viewed (a) as a set in the metric space $\mathbb{N}$; and (b) when viewed as a set in the metric space $\mathbb{R}$?

5. Show that a countable union of first category sets, in a given metric space X, is first category.

6. Show that a subset of a first category set is a first category set.

7. Let X be a complete metric space and suppose that $(F_k)_{k=1}^{\infty}$ is a family of closed sets in X, such that $X = \bigcup_{k=1}^{\infty} F_k$. Show that at least one of the sets F_k has an interior point.

8. Show that a set A in a metric space X is nowhere dense if and only if its complement A^c includes an open dense set.

9. Recall that x is said to be an isolated point of a metric space if the set $\{x\}$ is open. Show that a complete metric space without isolated points is uncountable. This gives another proof that $\mathbb{R}$ is uncountable, and that Cantor's middle thirds set is uncountable.

10. Show that in a complete metric space the intersection of a countable family of dense open sets is dense.

11. Show that a Baire space is uncountable.

12. Show that in a complete metric space, the intersection of countably many open sets, if non-empty and viewed as a metric space in its own right, is a Baire space.

Hint. See 3.2 exercise 8(b).

13. The complement of a first category set in a complete metric space is sometimes called a residual set. Let X be a complete metric space. Show that a countable intersection of residual sets in X is residual.

14. Let $(q_n)_{n=1}^{\infty}$ be an enumeration of the rationals. For each $\varepsilon > 0$ define the set $A_\varepsilon \subset \mathbb{R}$ as:

$$A_\varepsilon = \bigcup_{n=1}^{\infty} \,]q_n - \varepsilon 2^{-n-1}, \, q_n + \varepsilon 2^{-n-1}[.$$

We have here an open set including all rationals, expressed as a union of a countable infinite family of open intervals. The sum of the lengths of all the intervals is ε. This construction, and the fact that ε can be arbitrarily small, is often used to show that the set of rationals has Lebesgue measure 0.

What can you say about the intersection

$$\bigcap_{\varepsilon>0} A_\varepsilon \ ?$$

It plainly includes all rationals. Does it contain anything else?

15. a) Show that there cannot exist a continuous function $f : \mathbb{R} \to \mathbb{R}$, such that the set of all its discontinuities is the set of all irrationals.

   Hint. See 2.3 exercise 18.

   b) Show, on the other hand, that for any countable set $A \subset \mathbb{R}$, there exists a function for which A is the set of all its discontinuities.

   c) Exhibit a function whose set of discontinuities is Cantor's middle thirds set.

16. Show that an infinite-dimensional Banach space X cannot have a countable basis. In other words show that every basis of X is uncountable.

   Hint. Suppose that the sequence $(u_k)_{k=1}^\infty$ is a basis of X. Then $X = \bigcup_{n=1}^\infty E_n$ where $E_n = \text{span}(u_1, \ldots, u_n)$.

   Note. By a basis of a vector space X we mean a linearly independent set of vectors that spans X.

17. Let X be a complete metric space, let Y be a metric space and let W be a set of continuous mappings from X to Y. Suppose that for each $x \in X$ the set $\{f(x) : f \in W\}$ is bounded in Y. Show that there exists a non-empty open set U in X, such that the set $\{f(x) : f \in W, \ x \in U\}$ is bounded.

   Hint. Fix $a \in Y$ and consider the sets

$$A_n := \bigcap_{f \in W} f^{-1}(B_n^-(a)), \quad (n = 1, 2, \ldots)$$

18. The result of the previous exercise has an important application to functional analysis. Let X be a complete normed space, let Y be a normed space and let W be a set of continuous linear mappings from X to Y. Suppose that for each $x \in X$ the set $\{f(x) : f \in W\}$ is bounded in Y.

   a) Show that the set $\{f(x) : f \in W, \ x \in B_1\}$ is bounded, where B_1 is the unit ball in X.

   b) Deduce that the set W is bounded in the normed space $L(X, Y)$.

   Note. This important conclusion is called the *principle of uniform boundedness* or the *Banach-Steinhaus theorem*.

19. The principle of uniform boundedness (previous exercise) is often the key to proving something that seems highly plausible, and yet intractable. Try this.

Suppose that $1 \leq p < \infty$. Suppose that $(\lambda_n)_{n=1}^{\infty}$ is a sequence of real numbers, such that the series $\sum_{n=1}^{\infty} \lambda_n a_n$ is convergent for all $(a_n)_{n=1}^{\infty} \in \ell^p$. Prove that $(\lambda_n)_{n=1}^{\infty} \in \ell^q$, where q is the exponent conjugate to p (that is, $(1/p) + (1/q) = 1$; or $q = \infty$ if $p = 1$).

Hint. Consider the sequence of linear functionals $(\Lambda_n)_{n=1}^{\infty}$ where

$$\Lambda_n(a) = \sum_{k=1}^{n} \lambda_k a_k, \quad (a = (a_k)_{k=1}^{\infty} \in \ell^p).$$

Refer to 3.3 exercise 7 to finish off.

6.2 (◊) Genericity

In some areas of mathematics one speaks of *generic objects* or of a property that is generic or occurs generically, or of a predicate that is generically true. A diverse series of examples will give the flavour of what is meant:

a) Two lines in a plane are generically not parallel.
b) Two lines in three dimensional space generically do not meet.
c) A differentiable function of one real variable has generically only simple zeros.
d) A differentiable function of one variable has generically only non-degenerate maxima and minima (that is, at a local maximum or minimum the second derivative is not 0).
e) A generic $n \times n$ complex matrix has only simple eigenvalues.
f) A generic real number is irrational.

Hopefully these examples impart the idea that a generic object is in some sense typical. We expect π to be irrational, though it is not trivial to prove this. If it had turned out that π was rational, some explanation for this apparent coincidence, beyond the actual proof, might have been called for.

The objects under study—pairs of lines, differentiable functions, and so on—are modelled as points in a set X. Two principal ways have been proposed to give substance to the notion that a predicate is *generically true*. They have one thing in common: if we have countably many predicates that are generically true, then the predicate that asserts that they hold simultaneously is also generically true.

A. *Probabilistic genericity.* The set X admits a probability measure. The predicate is *almost certainly true*.

Recall that a measure on X is a σ-additive function $m : \mathcal{F} \to [0, \infty]$, where $\mathcal{F}$ is a σ-algebra of subsets of X. For the definitions of these notions see section 2.6.

A probability measure is a measure for which $m(X) = 1$. The point of a probability measure is to axiomatise the notion of probability. For a given set $A \in \mathcal{F}$ we interpret $m(A)$, which must lie in the interval $[0, 1]$ because $m(X) = 1$, as the probability that a randomly chosen point of X lies in A. The fact that $m(X) = 1$ says that it is certain that a random point of X lies in X. That seems quite sensible. But it is quite possible that a set $A \in \mathcal{F}$ satisfies $m(A) = 1$, although A is a proper subset of X. Similarly we can have $m(A) = 0$ although $A \neq \emptyset$. In the first case a random point x is "almost certainly" in A; in the second it is "almost certainly" not in A.

Now we can interpret the notion of genericity in probabilistic terms. Suppose we have a probability measure on X. A predicate applicable to elements of the set X is generically true if the set of elements for which it is true has measure 1. Equivalently the set of elements for which it is untrue has measure 0. It is almost certainly true.

B. *Topological genericity.* The set X is equipped with a metric d which makes it a complete metric space. The predicate is true in the complement of a first category set, in what is often called a *residual set*.

The idea that a nowhere dense set is somehow negligible is here extended to a first category set, that is, a countable union of nowhere dense sets. The rationals form a first category set in $\mathbb{R}$ so they are in this sense atypical of real numbers. This makes a certain amount of sense: there has to be some reason why a number, apparently randomly chosen, has a repeating decimal expansion, a coincidence that seems atypical. It also shows that density alone, a property shared by the rationals and the irrationals, is not enough to model typicality.

We illustrate these ideas by studying the property of differentiable functions, very plausibly generic, of having only simple zeros. We choose to work with the space $C^1(\mathbb{R})$. A metric for this space, making it a complete metric space, was described in section 3.3, and will be used here. We are concerned with the second version of genericity, topological genericity.

The property that f, a function in $C^1(\mathbb{R})$, has only simple zeros, is expressed by requiring that there exists no $x \in \mathbb{R}$, such that $f(x) = f'(x) = 0$.

Proposition 6.4 *The subset M of $C^1(\mathbb{R})$, consisting of all functions that have only simple zeros, is the complement of a first category set.*

Proof Let M be the set of all functions f in $C^1(\mathbb{R})$ that have only simple zeros. We then have $M = \bigcap_{N=1}^{\infty} L_N$, where

$$L_N = \left\{ f \in C^1(\mathbb{R}) : (f(s), f'(s)) \neq (0, 0) \text{ for all } s \in [-N, N] \right\}.$$

We shall show that L_N is open and dense in $C^1(\mathbb{R})$. From this it follows that M, though not open, is the complement of a first category set.

First, we prove that L_N is open. Let $f \in L_N$. Then the plane parametric curve

$$x = f(s), \quad y = f'(s)$$

does not pass through $(0, 0)$ in the interval $-N \le s \le N$. The set

$$C = \{(f(s), f'(s)) : -N \le s \le N\}$$

is compact and does not contain $(0, 0)$. Hence there exists $\varepsilon > 0$, such that the plane set

$$C_\varepsilon := \left\{ (u, v) \in \mathbb{R}^2 : |u - x| < \varepsilon, \ |v - y| < \varepsilon, \ (x, y) \in C \right\}$$

does not contain $(0, 0)$. Hence any function $g \in C^1(\mathbb{R})$ that satisfies

$$\sup_{[-N,N]} |g - f| < \varepsilon, \qquad \sup_{[-N,N]} |g' - f'| < \varepsilon$$

will have only simple zeros in the interval $-N \le s \le N$. Since the two displayed inequalities together define a neighbourhood of f in $C^1(\mathbb{R})$, we conclude that L_N is open.

As a preliminary to proving that L_N is dense in $C^1(\mathbb{R})$, we shall show that polynomials are dense in $C^1(\mathbb{R})$. The reader should recall the metric on $C^1(\mathbb{R})$, defined in section 3.3.

Let $\varepsilon > 0$. Choose m so that $2^{-m} < \varepsilon/4$. Then for any two functions f and g in $C^1(\mathbb{R})$ we have

$$\sum_{n=m+1}^{\infty} 2^{-n} \min \left(\sup_{[-n,n]} |f - g|, 1 \right) + \sum_{n=m+1}^{\infty} 2^{-n} \min \left(\sup_{[-n,n]} |f' - g'|, 1 \right) < \frac{\varepsilon}{2}.$$

Now

$$\sum_{n=1}^{m} 2^{-n} \min \left(\sup_{[-n,n]} |f - g|, 1 \right) + \sum_{n=1}^{m} 2^{-n} \min \left(\sup_{[-n,n]} |f' - g'|, 1 \right)$$

$$< \sup_{[-m,m]} |f - g| + \sup_{[-m,m]} |f' - g'|.$$

Hence to approximate a given $f \in C^1(\mathbb{R})$ by a polynomial with error less than ε we only have to find a polynomial g such that

$$\sup_{[-m,m]} |f - g| + \sup_{[-m,m]} |f' - g'| < \frac{\varepsilon}{2}.$$

This is possible by the Weierstrass approximation theorem (proposition 5.5). The required polynomial is an antiderivative of a polynomial that approximates f' uniformly and sufficiently closely on the interval $[-m, m]$.

Now we can prove that L_N is dense in $C^1(\mathbb{R})$. Let $f \in C^1(\mathbb{R})$ and let $\varepsilon > 0$. Approximate f in $C^1(\mathbb{R})$ by a polynomial g with error less than $\varepsilon/2$ in the metric of $C^1(\mathbb{R})$. Using the factorisation of g into first and second degree real polynomials (a consequence of the fundamental theorem of algebra) it is easy to see that a small change in the coefficients of g can be made, that changes it, with a corresponding change in the metric of less than $\varepsilon/2$, to a polynomial with simple roots only. Such a polynomial is in L_N (for any N actually) and approximates f with error less than ε in the metric of $C^1(\mathbb{R})$.   □

Of course the argument of the last paragraph shows directly that M is dense in $C^1(\mathbb{R})$, because every $f \in C^1(\mathbb{R})$ can be approximated by a polynomial with simple roots, but the fact that L_N is both open and dense for each N leads to the much stronger conclusion that M is a residual set.

6.2.1 Exercises

1. Show that a generic real number in the interval $[0, 1]$ is irrational, both in the topological sense, and in the probabilistic sense, using Lebesgue measure as a probability measure on $[0, 1]$.
2. Show that a generic real number in the interval $[0, 1]$ falls outside Cantor's middle thirds set, both in the topological sense, and in the probabilistic sense, using Lebesgue measure as a probability measure on $[0, 1]$.
3. Let X be the space of $n \times n$ complex matrices, with a suitable norm making it an n^2-dimensional Banach space. Let $A \subset X$ be the set of those matrices having only simple eigenvalues. Show that A is open and dense in X. Deduce that the set of all $n \times n$ matrices having a multiple eigenvalue is nowhere dense, and the property of having only simple eigenvalues is generic.

Hints. Some inputs from algebra are needed, as follows:

(i) *A is open.* A complex number λ is an eigenvalue of a matrix T with multiplicity r when it is a root of the characteristic polynomial

$$f(x) := \det(xI - T)$$

with multiplicity r. A polynomial $f(x) = x^n + a_{n-1}x^{n-1} + \cdots + a_0$ has a multiple root if and only if its discriminant D is 0. The latter is usually defined to be the product

$$\prod_{j \neq k} (\lambda_j - \lambda_k)$$

up to some normalising constant, ± 1. Here the sequence $\lambda_1, \ldots, \lambda_n$ comprises the roots of $f(x)$ counted with multiplicity. The point is that, by algebra, D is a polynomial function of the coefficients $a_0, a_1, \ldots, a_{n-1}$, and therefore, a polynomial in the entries of the matrix T.

(ii) *A is dense*. There is an invertible matrix C such that $C^{-1}TC$ is upper triangular.

Exercises 1 and 2 exhibit cases when topological genericity and probabilistic genericity coincide. Obviously we can destroy this coincidence by using a different probability measure, for example one concentrated on the rationals.

4. Let $(q_n)_{n=1}^{\infty}$ be an enumeration of the rationals in the interval $[0, 1]$. Define

$$\mu(A) := \sum_{q_n \in A} 2^{-n}$$

for each $A \subset [0, 1]$. Show that this is a probability measure on $[0, 1]$, such that $\mu(A) = 0$ if and only if A contains only irrationals.

It is less plausible that topological genericity and probabilistic genericity can diverge even using Lebesgue measure. In fact we shall describe, with the reader's help through a series of exercises, a *nowhere dense subset of* $[0, 1]$ *with positive Lebesgue measure*. In fact for any positive $\lambda < 1$ we can find an example with measure λ .

The procedure is to imitate the construction of Cantor's middle thirds set. Recall that this is a nowhere dense set in $[0, 1]$ with measure 0. It is built in an infinite sequence of steps by removing at the nth step one third of the set remaining at the $n - 1$st step. Now we want to remove a *variable fraction* r_n of the remaining set, in an effort to keep the same topological structure, but achieve at the end a positive measure. The use of the term 'fraction' implies that $0 < r_n < 1$.

At the first step we remove an open interval of length r_1 from the middle of the unit interval $[0, 1]$, leaving two disjoint closed intervals of equal length $(1 - r_1)/2$ and total length $s_1 = 1 - r_1$. After the nth step we have 2^n pairwise disjoint closed intervals, of equal length $2^{-n}s_n$ and total length s_n. We remove a fraction r_{n+1} of each remaining interval, taking from the middle of each interval, leaving 2^{n+1} pairwise disjoint closed intervals, of equal length $2^{-n-1}s_{n+1}$, and total length s_{n+1}.

5. a) Show that $s_n = \prod_{k=1}^{n}(1 - r_k)$.
   b) Let $0 \leq \lambda < 1$. Let $r_n = 1 - (\sigma_n/\sigma_{n-1})$ where σ_n is a strictly decreasing sequence in $[0, 1]$ such that $\sigma_0 = 1$ and $\lim_{n \to \infty} \sigma_n = \lambda$. Show that $\lim_{n \to \infty} s_n = \lambda$.
   c) For this item a minimum knowledge of measure theory is needed. Let $B_0 = [0, 1]$ and let B_n be the subset of $[0, 1]$ remaining after the nth step, with r_n defined as in item (b). Show that

$$m\left(\bigcap_{n=0}^{\infty} B_n\right) = \lambda$$

where m denotes Lebesgue measure.

d) Cantor's middle thirds set, corresponding to $r_n = 1/3$, is obtained by setting $\sigma_n = (2/3)^n$. Show that all cases when r_n is constant (they are the cases for which σ_n is in geometric progression) lead to $\lim_{n\to\infty} s_n = 0$, and that a necessary, but not sufficient, condition that s_n has a positive limit is that $r_n \to 0$.

e) Let

$$C = \bigcap_{n=0}^{\infty} B_n.$$

Show that C is closed and has empty interior.

Hint. If $]u, v[\subset \bigcap_{n=0}^{\infty} B_n$ a problem arises, because when n is large enough the constituent intervals of B_n become too short for $]u, v[$ to lie entirely within one of them.

The set C defined in exercise 5(e) is topologically the same as the Cantor set $2^{\mathbb{N}+}$, a fact that seems remarkable when you reflect that its Lebesgue measure can be 0.999. We can label the intervals of B_n with 'L' (left) or 'R' (right) as illustrated in Fig. 6.1 (sometimes a picture is worth a thousand words). To each element x of C we assign a sequence in the product $\{L, R\}^{\mathbb{N}+}$, such that the nth term of the assigned sequence is 'L' or 'R', according to whether x lies an interval of B_n marked 'L' or an interval of B_n marked 'R'. This defines a mapping

$$\phi : C \to \{L, R\}^{\mathbb{N}+}$$

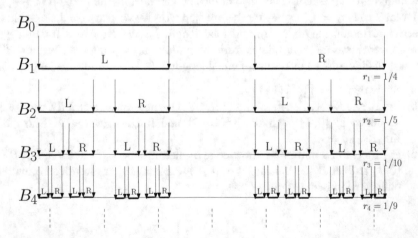

Fig. 6.1 A Cantor set with positive measure

6. Prove that the mapping ϕ is a homeomorphism of C on to $\{L, R\}^{\mathbb{N}+}$. The following steps can be used:

a) ϕ is bijective.

   Hint. Use Cantor's nested intersection theorem, proposition 6.1.
b) ϕ is continuous.
c) ϕ^{-1} is continuous.

   Hint. Proposition 4.10 can shorten the work, since both spaces are compact.

6.2.2   Pointers to Further Study

$\rightarrow$ Algebraic geometry.
$\rightarrow$ Sard's theorem.
$\rightarrow$ Liouville numbers.

6.3   (◇) Nowhere Differentiability

In 1872 Weierstrass gave an example of a continuous function $f : \mathbb{R} \rightarrow \mathbb{R}$ that is nowhere differentiable. The function in question was

$$f(x) = \sum_{n=1}^{\infty} a^n \cos(b^n \pi x)$$

where a satisfies $0 < a < 1$ and b is an odd integer satisfying $ab > 1 + \frac{3}{2}\pi$. We may assume that this example occasioned some surprise. In exercise 2 below is an example that is easier to analyse.

In this section we show that nowhere differentiable continuous functions are by no means rare. On the contrary they form the complement of a set of the first category in the space of continuous functions. The argument will demonstrate the existence of many such functions without indicating how to construct one.

The story of these functions is analogous to the discovery of transcendental numbers. No transcendental number was known to exist until Liouville exhibited an example of one, but they were then shown to exist in overwhelming abundance by a non-constructive argument of Cantor, showing that the set of all transcendental numbers is the complement of a countable set, and therefore has cardinality **c**.

Proposition 6.5 (Banach-Mazurkiewicz) *The set of all elements of the Banach space $C[0, 1]$, that are right differentiable nowhere in the interval $[0, 1[$, is the complement of a first category set.*

Proof We recall that the right derivative at a is the limit

$$D_r f(a) := \lim_{x \to a+} \frac{f(x) - f(a)}{x - a}$$

and that f is right differentiable at a when this limit exists and is finite. Therefore if f is an element of $C[0, 1]$, $0 \leq a < 1$ and f is right differentiable at a, then there exists $k > 0$, such that

$$- k(x - a) \leq f(x) - f(a) \leq k(x - a), \quad \text{for } a < x < 1. \tag{6.1}$$

In fact there exists k, such that $a < 1 - \frac{1}{k}$ and the inequalities hold. We only have to increase k if necessary. Geometrically (6.1) says that the slope of the chord joining $(a, f(a))$ to $(x, f(x))$ in the graph of f is less than k in absolute value.

For each positive integer k we define the set L_k as the set of all elements of $C[0, 1]$, such that there exists a in the interval $[0, 1 - \frac{1}{k}]$ for which the inequalities (6.1) hold (for $a < x < 1$). The set of all elements of $f \in C[0, 1]$, such that there exists $a \in [0, 1[$ at which f has a finite right derivative, is therefore a subset of the union $\bigcup_{k=1}^{\infty} L_k$. It is our object to show that each set L_k is closed and its complement dense in $C[0, 1]$. From this it will follow that the union is a first category set.

Proof that L_k is closed. Suppose that $(f_j)_{j=1}^{\infty}$ is a sequence in L_k that converges to a function g. This means of course that $f_j \to g$ uniformly in $[0, 1]$. We wish to show that $g \in L_k$. For each j there exists a point b_j in the interval $[0, 1 - \frac{1}{k}]$, such that inequalities (6.1) hold with f_j replacing f and b_j replacing a. A subsequence $(b_{n_j})_{j=1}^{\infty}$ converges to a point $c \in [0, 1 - \frac{1}{k}]$ and we have that $f_{n_j} \to g$ uniformly. Let x satisfy $c < x < 1$. Then for sufficiently large j we have $b_{n_j} < x$ and

$$-k(x - b_{n_j}) \leq f_{n_j}(x) - f_{n_j}(b_{n_j}) \leq k(x - b_{n_j}).$$

Letting $j \to \infty$ we deduce

$$-k(x - c) \leq g(x) - g(c) \leq k(x - c).$$

Hence $g \in L_k$.

Proof that the complement of L_k is dense. First we note that the space of all piece-wise affine continuous functions is dense in $C[0, 1]$. In fact, by the small oscillation theorem of fundamental analysis, given $f \in C[0, 1]$ and $\varepsilon > 0$, there exists a partition $0 = t_0 < t_1 < \cdots < t_m = 1$, such that the oscillation of f in each subinterval $[t_j, t_{j+1}]$ (that is, its maximum minus its minimum) is less than ε. Now all we have to do is join the points $(t_j, f(t_j))$ and $(t_{j+1}, f(t_{j+1}))$ by a straight line segment for each j, to obtain the graph of a piece-wise affine continuous function that approximates f uniformly with error less than ε.

Next we approximate the piece-wise affine continuous function with another piece-wise affine continuous function, which lies outside L_k, again with error less

Fig. 6.2 Uniform
approximation by a sawtooth
function

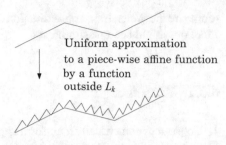

Uniform approximation
to a piece-wise affine function
by a function
outside L_k

than ε. We replace any straight segment of the graph with slope numerically less than or equal to k, by a saw-tooth pattern, where the teeth have small height (less than ε) and slope $\pm(k+1)$. This is illustrated in Fig. 6.2.

It follows that any function $f \in C[0, 1]$ can be uniformly approximated by a continuous function outside L_k; indeed by one that has at every point in $]0, 1]$ a left derivative with absolute value at least $k + 1$ and at every point in $[0, 1[$ a right derivative with absolute value at least $k + 1$. This shows that the complement of L_k is dense in $C[0, 1]$, or equivalently, since L_k is closed, that L_k is nowhere dense.

Finally therefore the set $\bigcup_{k=1}^{\infty} L_k$ is first category, and its complement is therefore second category. The complement comprises continuous functions that are right-differentiable nowhere in $[0, 1[$.   □

Functions having the properties of Weierstrass's function are, it appears, very plentiful. Are we prepared to say that the generic, or typical, continuous function has a finite derivative nowhere? In spite of the perfectly clear conclusion of proposition 6.5 an affirmative answer here seems unnatural, and it is interesting to enquire as to the reason for this. It may be because we are modelling an intuitive, even subjective idea, that of typicality or the absence of coincidence, by an objective mathematical idea, that of residuality (as the property of belonging to the complement of a first category set in a complete metric space is sometimes called). Could it be that residuality is an inappropriate way to model typicality?

It seems obvious that a typical differentiable function has only simple zeros; a tangency of the graph with the x-axis is such a coincidence. It is readily plausible that a typical real number is irrational; but we have to take a particular view, a digital approach to numbers, such as thinking of a number in terms of its decimal representation, in which case having only a finite number of non-zero digits (or being eventually periodic) is a major coincidence. A more naive view would suggest that, since we normally only operate with rational numbers, irrational numbers are a mere curiosity.

Similarly, when it comes to producing examples of elements of the space $C[0, 1]$, the bias towards differentiability seems inescapable. A sketch? Will have lots of derivatives. A formula? Usually lots of derivatives—it takes an effort to produce a function without derivatives. However, if one takes a digital approach to continuous functions, by systematically selecting approximations from a countable palette (compare the Weierstrass approximation theorem), then the scarcity of differentiable functions might become significant.

6.3.1   Exercises

1. Construct geometrically a uniform approximation to an affine function by a saw-tooth function with teeth having slope $\pm(k + 1)$ and differing from the initial function by less that ε.
2. Let $\{x\}$ denote the distance from the real number x to the nearest integer. Obviously the function $\{x\}$ is continuous, piece-wise affine, and periodic with period 1. Now define the function

$$f(x) = \sum_{n=1}^{\infty} 10^{-n}\{10^n x\}.$$

a) Show that f is continuous.

Let $0 \le x < 1$ and suppose that

$$x = 0.a_1 a_2 a_3 \ldots = \sum_{n=1}^{\infty} a_n 10^{-n}$$

with decimal digits a_k (in the range $0,\ldots,9$). Let $h_m = 10^{-m}$ if $a_m \ne 4$ or 9, but let $h_m = -10^{-m}$ if $a_m = 4$ or 9.

b) Show that $\{10^n(x + h_m)\} = \{10^n x\}$ if $n \ge m$.
c) Show that the difference quotient

$$\frac{f(x + h_m) - f(x)}{h_m}$$

is the sum of $m - 1$ numbers, each of which is ± 1.
d) Show that the limit

$$\lim_{m \to \infty} \frac{f(x + h_m) - f(x)}{h_m},$$

if it exists, cannot be a finite number, and deduce that f is nowhere differentiable.

6.3.2   Pointers to Further Study

$\rightarrow$ Divergent Fourier series.

6.4   Fixed Points

A fixed point of a mapping $f : X \rightarrow X$ is a solution to the equation $f(x) = x$. Fixed point theorems, asserting the existence of fixed points for mappings and spaces that satisfy certain conditions, are useful tools for proving the existence of desired objects in analysis. These are most often solutions to equations of various kinds.

Proposition 6.6 (Banach's Fixed Point Theorem[1]) *Let (X, d) be a complete metric space. Let $f : X \rightarrow X$ and suppose that there exists a number K in the interval $0 \leq K < 1$, such that $d(f(x), f(x')) \leq Kd(x, x')$ for all x and x' in X. Then there exists a unique $y \in X$ that satisfies $f(y) = y$.*

Proof Let $a_0 \in X$ and define the sequence $(a_n)_{n=0}^{\infty}$ iteratively by

$$a_{n+1} = f(a_n), \quad (n = 0, 1, 2, \ldots)$$

For $m < n$ we have, by the triangle inequality:

$$d(a_m, a_n) \leq d(a_m, a_{m+1}) + d(a_{m+1}, a_{m+2}) + \cdots + d(a_{n-1}, a_n)$$

$$\leq (K^m + K^{m+1} + \cdots + K^{n-1})d(a_0, a_1).$$

The series $\sum_{j=0}^{\infty} K^j$ is convergent since $0 < K < 1$. For all $\varepsilon > 0$ there exists N, such that if $N \leq m < n$ then

$$K^m + K^{m+1} + \cdots + K^{n-1} < \varepsilon.$$

But then we also have $d(a_m, a_n) < \varepsilon d(a_0, a_1)$. We conclude that $(a_n)_{n=0}^{\infty}$ is a Cauchy sequence and hence is convergent. Let its limit be y.

Now $f(a_n) = a_{n+1}$ and f is continuous. We let $n \rightarrow \infty$ and conclude that $f(y) = y$.

Finally we show that the equation $f(x) = x$ has only one solution. If a second solution $y' \neq y$ exists, then $d(y, y') \neq 0$, and

$$d(y, y') = d(f(y), f(y')) \leq Kd(y, y') < d(y, y').$$

This is a clear contradiction.   □

---

[1] Also called the contraction mapping principle.

We list some points of interest about the preceding proposition and fixed point theorems in general. Some topics are mentioned that the reader may encounter in further studies. Some of the claims are explored in the exercises.

a) It is important to observe that $K < 1$. The result does not necessarily hold if $K = 1$. Even assuming that $d(f(x), f(x')) < d(x, x')$ (for all $x \neq x'$) is not enough, though this is partially remedied if X is compact (exercises 2 and 3 below).

b) We shall call f a *strict contraction* if it satisfies the conditions of the proposition, and a *contraction* if it satisfies the weaker condition that $d(f(x), f(x')) < d(x, x')$ for all $x \neq x'$ (in other words a contraction reduces the distance between two distinct points, which seems logical enough). The terminology applied to these two cases varies somewhat in the literature and can be confusing. Perhaps a more logical terminology is to call the first a uniform contraction.

c) It follows from the proof that the solution may be calculated by the iteration $a_{n+1} = f(a_n)$, and, what is more, the starting point a_0 is completely arbitrary.

d) If the only condition on f is continuity then a fixed point does not necessarily exist. However for certain spaces a fixed point may exist for all continuous f. A famous example is $X = B_1^-(0)$ in $\mathbb{R}^n$. This is Brouwer's fixed point theorem. The existence of a fixed point for every continuous $f : X \to X$ is plainly a topological property of X.

e) Brouwer's theorem does not hold for continuous mappings $f : B_1^-(0) \to B_1^-(0)$ where $B_1^-(0)$ is the closed unit ball of an infinite-dimensional Banach space (such as $C[a, b]$ or ℓ^2). To ensure that a continuous mapping $f : B_1^-(0) \to B_1^-(0)$ has a fixed point in the case of an infinite-dimensional Banach space a further condition is needed. One such is that the range of f has compact closure (Schauder's fixed point theorem).

Some examples of important results in analysis that can be proved using fixed point theorems are:

f) The implicit function theorem in $\mathbb{R}^n$ and in Banach spaces.

g) Convergence of iteration methods to solve an equation $f(x) = 0$ for a given mapping $f : X \to X$ where $X = \mathbb{R}^n$ or a Banach space (see section 6.6).

h) The local existence of solutions to ordinary differential equations (see section 6.5).

i) The existence of solutions to some non-linear partial differential equations.

j) The existence of equilibrium strategies in game theory (Von Neumann, Nash).

6.4.1 Exercises

1. Find an example of metric spaces X and Y, a continuous mapping $f : X \to Y$, and a Cauchy sequence $(a_n)_{n=1}^\infty$ in X, such that $(f(a_n))_{n=1}^\infty$ is not a Cauchy sequence. Is this possible if X is complete?

2. Give an example of a complete metric space X and a mapping $f : X \to X$, such that $d(f(x), f(x')) < d(x, x')$ for all x, x' in X (except when $x = x'$), but f has no fixed point.

   Hint. Think of f as a graph.

3. Let X be a compact space, and let $f : X \to X$ satisfy $d(f(x), f(x')) < d(x, x')$ for all x and x', except when $x = x'$. Show that f has a unique fixed point.

   Hint. Study the function $g(x) := d(f(x), x)$.

4. Let (X, d) be a complete metric space and let $f : X \to X$ be continuous. Assume that there exists a positive integer n, such that f^n is a strict contraction. Show that f has a unique fixed point.

5. Let K be a real constant and let $T : C[0, 1] \to C[0, 1]$ be defined by

$$(Tu)(x) = K + \int_0^x u(t)\, dt, \quad (x \in [0, 1], \ u \in C[0, 1]).$$

Show that T^2 is a strict contraction, whereas T is not even a contraction according to the definition in point of interest (b) above. Find the unique fixed point of T (which exists by the previous exercise).

Banach's fixed point theorem requires a complete metric space. In applications one often works in a subset of a Banach space E. A closed ball in E is a complete metric space, which could be used as a candidate metric space for an application of proposition 6.6. However, it is often more convenient or desirable to work in an *open* ball. The following exercises address this question.

6. Let E be a Banach space, let U be the open ball with centre 0 and radius r. Let $f : U \to E$ have the following properties:

   i) There exists K in the interval $0 \leq K < 1$, such that

$$\|f(x_1) - f(x_2)\| \leq K \|x_1 - x_2\|$$

   for all x_1 and x_2 in U.

   ii) $\|f(0)\| < r(1 - K)$.

   Show that the mapping f has a unique fixed point in U. You can use the following steps:

   a) Show that $f(U) \subset U$.
   b) Show that the sequence $(f^n(0))_{n=0}^{\infty}$ converges to a point $y \in E$.

      Hint. Follow the proof of proposition 6.6.

   c) Show that

$$y = \sum_{n=0}^{\infty} \left(f^{n+1}(0) - f^n(0) \right).$$

d) Show that $y \in U$.

e) Show that y is the unique fixed point of f.

1. Let X and E be Banach spaces, let B be an open ball in X with centre 0 and let U be the open ball in E with centre 0 and radius r. Let $f : B \times U \to E$ satisfy the following conditions:

   i) There exists K in the interval $0 \le K < 1$ such that

$$\| f(x, u_1) - f(x, u_2) \| \le K \| u_1 - u_2 \|$$

   for all $x \in B$, $u_1 \in U$ and $u_2 \in U$.

   ii) $\| f(x, 0) \| < r(1 - K)$ for all $x \in B$.

   Show that there exists a unique mapping $g : B \to U$, such that

$$g(x) = f(x, g(x))$$

   for all $x \in B$. Show furthermore that g is continuous.

   Hint. Nearly all of this follows from the previous exercise. Only the continuity of g requires a new argument.

   Note. We have here the outline of the implicit function theorem of calculus. A proof of this theorem, for $X = \mathbb{R}^m$ and $E = \mathbb{R}^n$ for example, follows essentially this exercise, but combines it with the trappings of multivariate differentiation to obtain the conditions (i) and (ii).

6.5   (◇) Picard

We are going to use Banach's fixed point theorem to obtain the Picard-Lindelöf theorem for the existence and uniqueness of solutions to an ordinary differential equation $y' = f(x, y)$. Actually we prove a slightly weaker version of it, leaving the usual version, as presented in most texts on differential equations, as an exercise.

   The strategy in this, and in many similar applications of the Banach fixed point theorem, is to interpret the problem as a fixed point problem of the form $F(u) = u$, where F is a mapping whose domain is a closed ball $X := B_r^-(a)$ in a Banach space E, such that $F(X) \subset X$ and F is a strict contraction in X. The fixed point of F in X so obtained then yields the desired solution.

   The setting up of the problem requires an often delicate balance common to most applications of this kind. The domain, that is the ball $B_r^-(a)$, should be chosen so that F maps it into itself. This may require that it is not too small. Again, the ball should be chosen so that F is a strict contraction. This may require that it is not too big. It may also help to tinker with the norm, as what is or is not a contraction depends on the norm.

Our present problem is to solve an ordinary differential equation with initial condition. This is the *initial value problem*:

$$y' = f(x, y), \quad (x_0 - h < x < x_0 + h),$$

$$y(x_0) = y_0.$$

This problem was studied in section 4.5, but here we will impose stronger conditions and obtain uniqueness of the solution as well as existence. The result, the Picard-Lindelöf theorem, is the basis for most of the theory of ordinary differential equations. It is not assumed that the reader has read 4.5.

We assume that $f : A \to \mathbb{R}^2$, that A is open in $\mathbb{R}^2$, the mapping f is continuous and $(x_0, y_0) \in A$. The aim is to show that a solution $y = \phi(x)$ exists on a sufficiently small interval $]x_0 - h, x_0 + h[$ centred at x_0, such that the initial condition $\phi(x_0) = y_0$ is satisfied. The radius h of the interval can also be considered at this point to be unknown. Note that when we say that ϕ is a solution it is implied that the graph of ϕ is a subset of A (or lies in A, as is natural to say here).

For simplicity we take the initial condition to be of the simplest kind, with $x_0 = y_0 = 0$. The general result as stated above can be obtained by translation of $(0, 0)$ to (x_0, y_0).

Instead of solving the problem as it stands we convert it into an equivalent *integral equation*

$$y(x) = \int_0^x f(t, y(t)) \, dt, \quad (-h \le x \le h).$$

Here we seek a continuous function $\phi(x)$ defined on the closed interval $[-h, h]$ that satisfies

$$\phi(x) = \int_0^x f(t, \phi(t)) \, dt, \quad (-h \le x \le h).$$

Switching from the open interval $]-h, h[$ to the closed interval $[-h, h]$ makes no difference since, in any case, h has still to be specified. It also means that we may employ the Banach space $C[-h, h]$. The reader should convince themselves that a continuous solution to the integral equation will automatically be differentiable and will satisfy the initial value problem in $]-h, h[$.

The next step is to convert the problem into a fixed point problem in a closed ball in the Banach space $C[-h, h]$. Assume that f is continuous in a rectangle D, the latter defined by $|x| \le a$, $|y| \le b$. Let $M = \sup_D |f|$.

Now we introduce a new restriction on f, making f a little more than just continuous.

The Lipschitz Condition. There exists K, such that

$$|f(x, y_1) - f(x, y_2)| \le K|y_1 - y_2|$$

for all (x, y_1), (x, y_2) in D.

We reduce a (if needed) so as to satisfy the two conditions:

$$aM \le b, \quad aK < 1. \tag{6.2}$$

We shall obtain a solution to the integral equation on the interval $[-a, a]$; so in fact we are letting $h = a$.

Let X be the metric space that consists of all continuous functions ϕ mapping $[-a, a]$ to $\mathbb{R}$, and that satisfy $|\phi| \le b$. This is just the closed ball $B_b^-(0)$ in the Banach space $C[-a, a]$ with the norm $\|f\|_\infty$, and is therefore complete.

For each $\phi \in X$ we define $T(\phi)$ as the function

$$T(\phi)(x) = \int_0^x f(t, \phi(t)) \, dt, \quad (-a \le x \le a)$$

In the first place the definition of T makes sense. For if $\phi \in X$ then $|\phi(t)| \le b$, and $f(t, \phi(t))$ is defined. The integral certainly exists since ϕ, and therefore $f(t, \phi(t))$, is continuous, and it defines a continuous function of its upper limit x.

In the second place T maps X into itself. As we have seen, the function $T(\phi)$ is continuous in $[-a, a]$. We also have

$$|T(\phi)(x)| \le \left| \int_0^x \left| f(t, \phi(t)) \right| dt \right| \le Ma \le b,$$

by the first condition in (6.2). That is, $T\phi \in X$.

In the third place T is a strict contraction. Consider two functions ϕ_1 and ϕ_2 in X. Then

$$|T(\phi_1)(x) - T(\phi_2)(x)| \le \left| \int_0^x \left| f(t, \phi_1(t)) - f(t, \phi_2(t)) \right| dt \right|$$

$$\le K \left| \int_0^x \left| \phi_1(t) - \phi_2(t) \right| dt \right|$$

$$\le Ka \|\phi_1 - \phi_2\|_\infty.$$

This says that

$$\|T(\phi_1) - T(\phi_2)\|_\infty \le Ka \|\phi_1 - \phi_2\|_\infty$$

and so, by the second condition of (6.2), the mapping T is a strict contraction.

Finally, we apply Banach's fixed point theorem to obtain a unique fixed point in X, thus also solving the initial value problem.

6.5.1   Exercises

1. The Picard-Lindelöf theorem as usually proved in courses on differential equations is stronger than the version proved above, in that the condition $aK < 1$ (the second condition in (6.2)) is not needed. Only the first condition, $aM \leq b$, is required. The interval of existence thus obtained does not therefore depend on the actual value of K. Prove this by modifying the norm in $C[-a, a]$ to the equivalent norm

$$\|f\|_\beta = \sup_{-a \leq x \leq a} |e^{-\beta|x|} f(x)|$$

where β is a positive constant, to be suitably determined so that Banach's fixed point theorem can be applied.
2. The same improvement as was stated in the previous exercise can also be obtained by applying the result of 6.4 exercise 4. You will have to express an iterated integral as a single integral.

6.6   (◊) Zeros

Suppose we have a function $f : \mathbb{R} \to \mathbb{R}$, and for a given $a \in \mathbb{R}$ the value $f(a)$ is small, but the derivative $f'(a)$ is big. May we not deduce that a root of $f(x) = 0$ lies near to a? It seems as if the graph of the function is forced across the x-axis in the vicinity of a.

In a similar way we could consider a mapping $f : \mathbb{R}^n \to \mathbb{R}^n$, and a point a such that $|f(a)|$ (the Euclidean norm) is small, but what could it mean for the derivative of f at a to be big?

It is assumed that the reader is familiar with the idea that the derivative $Df(a)$ of a mapping $f : \mathbb{R}^n \to \mathbb{R}^n$ at a point a is a linear operator from $\mathbb{R}^n$ to $\mathbb{R}^n$. The derivative may then be identified with an $n \times n$-matrix. Now there are several ways to assign a norm to matrices. This will be discussed in a moment. Assuming this is done it is easier to require that the inverse of $Df(a)$ is small. That is, that $\|Df(a)^{-1}\|$ is small, where we employ here a matrix norm. So the question is this: given that $|f(a)|$ is small and $\|Df(a)^{-1}\|$ also small, is there a solution of $f(x) = 0$ near to a? And how small should the first-mentioned quantities be?

If we use the Euclidean norm $|v|$ for vectors then a natural norm for a matrix T is the operator norm defined in 2.4. This is given by

$$\|T\| = \max_{|v| \le 1} |Tv|.$$

It is equivalent to saying that $\|T\|$ is the lowest number K, such that $|Tv| \le K|v|$ for all vectors v.

Unfortunately it is difficult to calculate this matrix norm. It can be shown to be the square-root of the highest eigenvalue of the matrix T^*T. For this reason we often use a different, and somewhat bigger norm, for example the Euclidean norm in $\mathbb{R}^{n^2}$, the quantity $(\sum_{ij} a_{ij}^2)^{1/2}$. However the matrix norm is chosen, it is essential that it satisfies $|Tu| \le \|T\| \, |u|$ for all vectors u.

We are going to prove a proposition that resembles Kantorovich's version of the Newton approximation method in n dimensions, but is somewhat weaker and is simpler to prove. It provides a nice example of the use of Banach's fixed point theorem while not requiring the more difficult estimates needed for Kantorovich's theorem. It also yields some quite useful information about the location of zeros of a vector-valued mapping.

Proposition 6.7 *Let $f : A \to \mathbb{R}^n$ be a differentiable mapping, where $A \subset \mathbb{R}^n$ is an open set. Assume that the derivative Df satisfies a Lipschitz condition*

$$\|Df(x) - Df(y)\| \le M|x - y|, \quad (x, y \in A).$$

Let $a \in A$, assume that $f(a) \ne 0$, that $Df(a)$ is invertible and

$$\|Df(a)^{-1}\| \, |Df(a)^{-1} f(a)| < \frac{1}{4M}.$$

Let

$$R_- = \frac{1 - \sqrt{1 - 4M\|Df(a)^{-1}\| \, |Df(a)^{-1} f(a)|}}{2M\|Df(a)^{-1}\|},$$

$$R_+ = \frac{1}{M\|Df(a)^{-1}\|}.$$

Then we have the following conclusions:

1. *If $B_{R_-}^-(a) \subset A$ then the equation $f(x) = 0$ has a unique solution in the ball $B_{R_-}^-(a)$.*
2. *If $R_- < R < R_+$ and $B_R^-(a) \subset A$ then the solution, declared to exist in the previous item, is the unique solution in the ball $B_R^-(a)$.*

Proof We shall prove the existence of a solution by applying Banach's fixed point theorem in the complete metric space $B_R^-(a)$, where, for the moment, we leave R

unspecified. Instead of solving the equation $f(x) = 0$ we will solve the fixed point problem $g(x) = x$, where

$$g(x) := x - Df(a)^{-1}f(x).$$

It is clear that the problem $g(x) = x$ is equivalent to $f(x) = 0$.

We begin by estimating the difference $g(x) - g(y)$, where x and y belong to the ball $B_R^-(a)$, which we suppose is included in A. We have

$$g(x) - g(y) = x - Df(a)^{-1}f(x) - y + Df(a)^{-1}f(y)$$
$$= -Df(a)^{-1}\big(f(x) - f(y) - Df(a)(x - y)\big)$$

leading to

$$|g(x) - g(y)| \leq \|Df(a)^{-1}\|\,\big|f(x) - f(y) - Df(a)(x - y)\big|.$$

We estimate the second factor using a version of the *mean value theorem* for multivariate calculus. Let $L(x, y)$ be the line segment joining x and y. Then (the mean value theorem is the inequality in the first line):

$$\big|f(x) - f(y) - Df(a)(x - y)\big| \leq \sup_{z \in L(x,y)} \|Df(z) - Df(a)\|\,|x - y|$$
$$\leq M \sup_{z \in L(x,y)} |z - a|\,|x - y|$$
$$\leq MR\,|x - y|.$$

It follows from this that for all x and y in $B_R^-(a)$ we have:

$$|g(x) - g(y)| \leq MR\|Df(a)^{-1}\|\,|x - y|.$$

By requiring that $MR\|Df(a)^{-1}\| < 1$ we can ensure that g is a strict contraction in $B_R^-(a)$, but we cannot use the Banach fixed point theorem while we cannot yet assert that g maps $B_R^-(a)$ into itself. However, the above estimate for $|g(x) - g(y)|$ does imply that if $R < 1/M\|Df(a)^{-1}\|$, and if $B_R(a) \subset A$, then a fixed point of g in $B_R(a)$ is unique, if one exists (the reader should check this, see exercise 1). This is conclusion 2.

If $x \in B_R^-(a)$ we have, using the above estimate for $|g(x) - g(y)|$:

$$|g(x) - a| \leq |g(x) - g(a)| + |g(a) - a|$$
$$\leq MR\|Df(a)^{-1}\|\,|x - a| + |Df(a)^{-1}f(a)|$$
$$\leq MR^2\|Df(a)^{-1}\| + |Df(a)^{-1}f(a)|$$

So to ensure both that g is a strict contraction, and that it maps $B_R^-(a)$ into itself, we need R to satisfy two inequalities:

i) $MR\|Df(a)^{-1}\| < 1$

ii) $MR^2\|Df(a)^{-1}\| + |Df(a)^{-1}f(a)| \le R$

Notice that (ii) implies (i), (since $f(a) \ne 0$; but see exercise 2 below). Condition (ii) is equivalent to requiring that R is either a root of, or lies between the roots of the quadratic polynomial

$$\left(M\|Df(a)^{-1}\|\right)X^2 - X + |Df(a)^{-1}f(a)|.$$

The lower root is precisely the number R_-.   □

The proof also shows that the solution can be computed by the *simplified Newton iteration*

$$x_{n+1} = x_n - Df(a)^{-1}f(x_n), \quad x_0 = a.$$

Kantorovich's theorem uses the weaker inequality

$$\|Df(a)^{-1}\| \, |Df(a)^{-1}f(a)| < \frac{1}{2M} \tag{6.3}$$

as its premise, but employs the more powerful Newton iteration

$$x_{n+1} = x_n - Df(x_n)^{-1}f(x_n), \quad x_0 = a,$$

with its characteristic accelerated convergence, in its proof.

Another point is that the proof holds *totally without change* in the context that $\mathbb{R}^n$ is replaced by a Banach space E, which may be infinite-dimensional. The derivatives $Df(x)$ are then defined as so-called Frechet[2] derivatives. To be precise, $Df(x)$ is a linear mapping from E to itself, with the property that, for every $\varepsilon > 0$ there exists $\delta > 0$, such that

$$\|f(y) - f(x) - Df(x)(y - x)\| \le \varepsilon\|y - x\|$$

for all y in the ball $B_\delta(x)$. This is a topic for 'Further studies'.

---

[2] M. R. Fréchet seems to have the first to study metric spaces, in his 1906 PhD dissertation.

6.6.1 Exercises

1. Let X be a metric space and let $A \subset X$. Suppose that $f : A \to X$ and that there exists K, such that $0 \leq K < 1$, and such that

$$d(f(x), f(y)) \leq K d(x, y)$$

for all x and y in A. Show that if a fixed point of f exists in A, then it is uniquely determined.

Note. This confirms a point used in the proof of proposition 6.7.

2. Let $f : A \to \mathbb{R}^n$ be a differentiable mapping, where $A \subset \mathbb{R}^n$ is an open set. Assume that the derivative Df satisfies a Lipschitz condition

$$\|Df(x) - Df(y)\| \leq M|x - y|, \quad (x, y \in A).$$

Let $a \in A$, assume that $f(a) = 0$, and that $Df(a)$ is invertible. Show that a is the only solution of $f(x) = 0$ in the open ball $B_R(a)$, where $R = 1/M\|Df(a)^{-1}\|$, provided $B_R(a) \subset A$.

Hint. Run through the proof of proposition 6.7, but assuming that $f(a) = 0$.

Note. This can be improved using Taylor's theorem, or else a new version of proposition 6.7. See below and exercise 5.

3. Let $f(z) := \sum_{k=0}^{m} a_k z^k$ be a polynomial in the complex variable z with complex coefficients. Let $\rho > 0$ and set

$$M = \sum_{k=2}^{m} k(k - 1)|a_k|\rho^{k-2}.$$

Suppose that

$$4|a_0|M < |a_1|^2,$$

and set

$$R_- := \frac{|a_1| - \sqrt{|a_1|^2 - 4M|a_0|}}{2M}, \quad R_+ = \frac{|a_1|}{M}.$$

Deduce the following consequences of proposition 6.7:

a) If $R_- \leq \rho$ the polynomial $f(z)$ has a unique root in the closed disc with centre 0 and radius R_-.

b) Let $R_- \leq R < \min(R_+, \rho)$. Then the root of $f(z)$, whose existence is asserted in item (a), is the unique root in the open disc with centre 0 and radius R.

Note. These results can be improved by using a new version of proposition 6.7. See below and exercise 5.

4. In the notation and with the premises of proposition 6.7, the solution y of $f(x) = 0$ in the ball $B_R^-(a)$ can be found by the iteration

$$x_{n+1} = x_n - Df(a)^{-1} f(x_n), \quad x_0 = a.$$

Let $K = M R_- \|Df(a)^{-1}\|$. Show that for all n we have the error estimate

$$\|y - x_n\| \le C K^n,$$

where C is a constant (that is C is independent of n).

Note. This has practical significance as it tells that, roughly speaking, for each unit increase in n, we can expect that the approximation acquires around the same number, $|\log_{10} K|$, of new correct decimal digits.

Without pursuing the full version of the Newton-Kantorovich theorem (which is under the heading of 'Pointers to further study') one can try to obtain a result based on its weaker premise (6.3), but still using the simplified Newton iteration of proposition 6.7.

Some more sophisticated multivariate calculus will be needed. In particular the estimate provided by the mean value theorem, and used in the proof of proposition 6.7, namely

$$\left| f(x) - f(y) - Df(a)(x - y) \right| \le M \sup_{z \in L(x,y)} |z - a|\,|x - y|, \tag{6.4}$$

can be improved when $y = a$ by a crucial factor of $\frac{1}{2}$, to:

$$\left| f(x) - f(a) - Df(a)(x - a) \right| \le \tfrac{1}{2} M |x - a|^2. \tag{6.5}$$

This is essentially a case of Taylor's theorem. Now we can state a version of proposition 6.7, based on the weaker premise (6.3). To compensate for the weaker premise the statement of the result is a bit more fussy:

Let $f : A \to \mathbb{R}^n$ be a differentiable mapping, where $A \subset \mathbb{R}^n$ is an open set. Assume that the derivative Df satisfies a Lipschitz condition

$$\|Df(x) - Df(y)\| \le M|x - y|, \quad (x, y \in A).$$

Let $a \in A$, assume that $f(a) \ne 0$, that $Df(a)$ is invertible and

$$\|Df(a)^{-1}\|\,|Df(a)^{-1} f(a)| < \frac{1}{2M}.$$

Let

$$b = a - Df(a)^{-1}f(a), \qquad r = |Df(a)^{-1}f(a)|$$

and assume that the ball $B_r^-(b)$ is entirely included in A. Then

1. *The equation $f(x) = 0$ has a unique solution in the ball $B_r^-(b)$.*
2. *The solution can be obtained by the simplified Newton iteration*

$$x_{n+1} = x_n - Df(a)^{-1}f(x_n), \qquad x_0 = a.$$

Kantorovich's theorem asserts that under the same conditions the Newton iteration

$$x_{n+1} = x_n - Df(x_n)^{-1}f(x_n)$$

converges to the solution. The price paid, in practical applications, for the accelerated convergence of the Newton iteration, is the need to recalculate the derivative at each step. The above proposition, using the simplified Newton iteration, might therefore have practical value. The reader can supply its proof using the steps laid out in the following exercise:

5. To simplify the notation we set $\beta := \|Df(a)^{-1}\|$. The assumption is therefore that $2M\beta r < 1$. We also let

$$g(x) := x - Df(a)^{-1}f(x), \qquad (x \in A),$$

as in the proof of proposition 6.7.

a) Show that

$$|g(x) - g(y)| \leq 2M\beta r|x - y|$$

for all x and y in $B_r^-(b)$.

Hint. Use (6.4).

b) Show that if $|x - b| \leq r$ then

$$|g(x) - b| \leq 2M\beta r^2.$$

Hint. This is where you will need (6.5).

c) Show that g is a strict contraction of $B_r^-(b)$ into itself and obtain the stated conclusion.

d) A further conclusion. The solution obtained to $f(x) = 0$ lies in the ball $B_{2r}^-(a)$. Show that if $2r < R < 1/M\beta$, and if $B_R^-(a) \subset A$, then the solution is unique in $B_R^-(a)$.

6. Show that the conclusion of the previous exercise cannot be improved upon, in the following sense. If we adopt the premise $M\beta r < \lambda$, where $\lambda > 1/2$, then it can happen that no R exists such that $f(x) = 0$ has a *unique* solution in the ball $B_R^-(a)$.

   Hint. Take f to be a second degree polynomial $f(z)$ of a complex variable, regarded as a mapping from $\mathbb{R}^2$ to itself. Compare exercise 3.

6.6.2   Pointers to Further Study

$\rightarrow$ Newton-Kantorovich theorem
$\rightarrow$ Frechet derivatives
$\rightarrow$ Implicit function theorem

6.7   Completion of a Metric Space

If the metric space (X, d) is not complete, then it is possible to enlarge X in order to build a complete metric space X^*. Furthermore this can be done in such a way that X is dense in X^*.

We circumvent the thorny question of where the new elements, that are joined to X to form X^*, should come from. Instead, we produce a complete space Y, and an isometry J of X on to a subspace $J(X)$ of Y. Then we define X^* to be the closure of $J(X)$ in Y. The idea is that two spaces that are isometric to each other, as are X and $J(X)$, can be thought of as identical metric spaces, even though they are built out of wildly different materials.

Proposition 6.8 *Let (X, d) be a metric space. Then there exists a complete metric space (X^*, d^*) and a mapping $J : X \to X^*$, such that J is an isometry of X on to $J(X)$, and $J(X)$ is dense in X^*.*

Proof Let $Y = C_B(X)$, that is, the space of all bounded, continuous real-valued functions on X, with the norm $\| f \|_\infty = \sup_{x \in X} |f(x)|$. We know that Y is complete (proposition 3.9, although X is not necessarily complete).

Fix a point $a \in X$. For each $u \in X$ we build the function $f_u : X \to \mathbb{R}$ by setting

$$f_u(x) = d(x, u) - d(x, a), \quad (x \in X).$$

For each u the function f_u is continuous. Furthermore

$$d(x, u) - d(x, a) \le d(u, a),$$

so that f_u is bounded. Hence $f_u \in Y$. We shall show that the mapping $J : X \to Y$, that assigns to each $u \in X$ the function f_u, is an isometry of X on to $J(X)$.

We want to show that $\| f_u - f_v \|_\infty = d(u, v)$. Let $u, v \in X$. Then

$$|f_u(x) - f_v(x)| = |d(x, u) - d(x, a) - d(x, v) + d(x, a)|$$

$$= |d(x, u) - d(x, v)| \le d(u, v)$$

and hence $\| f_u - f_v \|_\infty \le d(u, v)$. Moreover $|f_u(u) - f_v(u)| = d(u, v)$, so that we must have $\| f_u - f_v \|_\infty = d(u, v)$. We conclude that $J : X \to Y$, defined so that $u \mapsto f_u$, is an isometry on to its range.

Finally we let $X^* = \overline{J(X)}$, the closure of the range of J in Y. The mapping J is then an isometry from X on to a dense subset of X^*, and since Y is complete and X^* a closed set, the latter is also complete.   □

Definition A completion of a metric space (X, d) is a complete metric space (X^*, d^*) together with an isometry of X on to a dense subset of X^*.

We have seen one example of a completion. In a precise sense all completions of a given metric space are the same.

Proposition 6.9 *Let (X, d) be a metric space. Suppose that there exist complete metric spaces (Y_1, d_1) and (Y_2, d_2), together with mappings $f_1 : X \to Y_1$ and $f_2 : X \to Y_2$, such that f_1 and f_2 are isometries on to their respective ranges, $f_1(X)$ is dense in Y_1 and $f_2(X)$ is dense in Y_2. Then there exists an isometry g of Y_1 on to Y_2, such that $f_2 = g \circ f_1$.*

Proof We define g first of all on the set $f_1(X)$, which is dense in Y_1, in such a way that g maps $f_1(x)$ to $f_2(x)$. It is clear that g is an isometry, and therefore has an extension to Y_1 by proposition 5.2. The extension is also an isometry, and we denote it also by g.

It remains only to show that g is surjective. Let $y \in Y_2$. Then there exists a Cauchy sequence $(x_n)_{n=1}^\infty$ in X, such that $y = \lim_{n \to \infty} f_2(x_n) = \lim_{n \to \infty} g(f_1(x_n))$. But then the sequence $(f_1(x_n))_{n=1}^\infty$ is a Cauchy sequence in Y_1, is therefore convergent, and $y = g(\lim_{n \to \infty} f_1(x_n)) \in g(Y_1)$.   □

The proof is illustrated in Fig. 6.3.

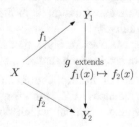

Fig. 6.3 Completions are unique

6.7.1   Other Ways to Complete a Metric Space

Another popular way to complete a metric space is to use the Cauchy sequences of X to build new elements. We leave it to the reader to supplement the following scanty description with the full details. Unfortunately the details are long, so that it may be best to look them up in a text that covers this approach.

Two Cauchy sequences $a = (a_n)_{n=1}^{\infty}$ and $b = (b_n)_{n=1}^{\infty}$ are said to be equivalent, symbolically $a \sim b$, if $\lim_{n \to \infty} d(a_n, b_n) = 0$. The relation $a \sim b$ is an equivalence relation on the set of all Cauchy sequences in X. The set X^* consists of the set of all equivalence classes of Cauchy sequences.

Denote the equivalence class of a Cauchy sequence a by $[a]$. We map X to X^* (thus identifying X with a subset of X^*) by setting $J(x) = [(x, x, x, \ldots)]$.

We define a metric on X^* by setting $d^*([a], [b]) := \lim_{n \to \infty} d(a_n, b_n)$. The limit here exists, because it may be shown that $(d(a_n, b_n))_{n=1}^{\infty}$ is a Cauchy sequence in $\mathbb{R}$. In order to show that the metric d^* is well-defined one must show that $d^*([a], [b])$, as defined, really only depends on the equivalence classes $[a]$ and $[b]$. In other words one must show that, if $[a] = [c]$ and $[b] = [d]$, then $\lim_{n \to \infty} d(a_n, b_n) = \lim_{n \to \infty} d(c_n, d_n)$.

Many details remain. One must show that d^* is a metric. The main challenge here is the triangle inequality. After this it still remains to show:

i) (X^*, d^*) is complete.
ii) J is an isometry.
iii) $J(X)$ is dense in X^*.

6.7.2   Exercises

1. Fill in the details of the completion by Cauchy sequences as outlined above; or look them up in another text.
2. Let (X, d) be a metric space and let A be a dense set in X. Show that the completions of the metric spaces (X, d) and (A, d) are isometric.
3. Show that the completion of (X, d) can be constructed by using the space $B(X)$ of bounded functions, instead of the space $C_B(X)$ of bounded continuous functions, following the same construction as in the text. In this way we avoid having to think about continuity.
4. Let X be a normed vector space over $\mathbb{R}$. Show that X^* can be given the structure of a complete normed vector space, such that X is a linear subspace of X^*.

   Hint. Define sum $u + v$, scalar multiplication λu, and norm $\|u\|$ in X^* by approximating u and v by sequences in X. Don't forget to show that the outcomes are well-defined, that is, independent of which approximating sequences are used.

Chapter 7
Connected Spaces

The most intuitive of topological properties should allow us to say that a space in one piece is not homeomorphic to a space in two pieces. You cannot pass continuously from one piece of string to two pieces; you have to cut it.

We have seen how a basic theorem of analysis, that a continuous function on the interval $[0, 1]$ attains a maximum and a minimum, is the germ of the notion of compactness. Another basic theorem, the intermediate value theorem, that the continuous image of an interval is also an interval, is the germ of the notion of connectedness.

7.1 Connectedness

Does there exist a homeomorphism from $B := [0, 1]$ on to the unit circle $T \subset \mathbb{R}^2$? Both spaces are compact—no obstacle there to the existence of a homeomorphism.

If we remove one point, say $\frac{1}{2}$, from B, the space "falls into two bits". It does not seem possible to remove a point from T with the effect that the space falls apart. In fact on removing one point from T we obtain a space homeomorphic to $Y :=]0, 1[$; see, for example, 2.5 exercise 8. Another way to see this is to identify $\mathbb{R}^2$ with $\mathbb{C}$ and use the mapping $f :]0, 1[\rightarrow T \setminus \{1\}$, given by $f(t) = e^{2\pi i t}$, which is a homeomorphism (see 4.2 exercise 6). It does not matter which point is removed; $T \setminus \{p\}$ is homeomorphic to $]0, 1[$.

The question initially posed can therefore be answered negatively if we can show: there is no homeomorphism

$$\phi : [0, \tfrac{1}{2}[\cup]\tfrac{1}{2}, 1] \rightarrow]0, 1[.$$

© The Author(s), under exclusive license to Springer Nature Switzerland AG 2022   219
R. Magnus, *Metric Spaces*, Springer Undergraduate Mathematics Series,
https://doi.org/10.1007/978-3-030-94946-4_7

The domain $[0, \frac{1}{2}[\cup]\frac{1}{2}, 1]$ has the property that $[0, \frac{1}{2}[$ and $]\frac{1}{2}, 1]$ are both open sets relative to it. It is the union of two open, disjoint and non-empty open sets. Note that these open sets are also closed, for each is the other's complement.

But what of the range $]0, 1[$? Is *it too* the union of two open, disjoint and non-empty open sets? We are going to show that this is not so, and introduce for this purpose a new and fascinating topological property.

Definition The metric space (X, d) is said to be *connected* if it is not the union of two open, disjoint and non-empty sets.

A space that is not connected is said to be disconnected. It is clear that connectedness is a topological property.

The definition of connectedness is phrased negatively; a certain kind of decomposition does not exist. This means that proofs by contradiction are frequent. It is also useful, as with nearly all important properties in metric space theory, to have a variety of paraphrases at one's disposal. Here are two that are often convenient:

a) The space (X, d) is connected if and only if the only sets that are both open and closed in X are X itself and the empty set $\emptyset$.
b) The space (X, d) is connected if and only if, whenever $X = A \cup B$, with A and B both open and $A \cap B = \emptyset$, then either A or B is empty.

Proposition 7.1 $\mathbb{R}$ *is connected.*

Proof Suppose that $\mathbb{R}$ is disconnected, that is, we suppose that $\mathbb{R} = A \cup B$, where A and B are open, disjoint, and neither is empty. Both sets are therefore closed.

Let $p \in B$. Either there exists $x \in A$ with $x < p$, or else there exists $x \in A$ with $x > p$. Assume the former case (the latter case is similar). Let

$$u = \sup\{x \in A : x < p\}.$$

Then $u \leq p$. Moreover, either u is in A or it is a limit point of A; therefore, A being closed, we must have $u \in A$. But then $u \notin B$ so that we must have $u < p$. But then $]u, p] \subset B$ by the definition of u, so that u must be a limit point of B. But since B is closed we then have $u \in B$ also. This is a contradiction since we assumed that $A \cap B = \emptyset$.   □

Note how we relied heavily on the existence of supremum and infimum, and therefore on the Dedekind completeness of the real number field. This was inevitable. The space $\mathbb{Q}$ of rationals is not connected, a fact the reader might enjoy proving now (but we shall soon identify all connected subspaces of $\mathbb{R}$). We can even regard the proposition that $\mathbb{R}$ is connected as yet another version of the axiom of completeness, although we would then have to view $\mathbb{R}$ initially as a topological space with the order topology rather than as a metric space (see section 2.6).

We know that $\mathbb{R}$ is homeomorphic to $]0, 1[$, and so the latter is connected. The space $T \setminus \{p\}$, obtained by removing one point from the circle, is therefore connected. We conclude that T is not homeomorphic to any kind of interval. For if one point, excepting an endpoint, is removed from an interval, the resulting space is disconnected.

7.1.1 Connected Sets

There is a standard way to transfer predicates applicable to metric spaces and apply them to sets in metric spaces.

Definition Let (X, d) be a metric space and let $M \subset X$. We say that M is a connected set if the metric space (M, d) is connected.

Literally this means that there is no decomposition $M = A \cup B$, where A and B are open *relative to* M, $A \cap B = \emptyset$, but neither is empty.

It is often troublesome to handle sets open relative to M. It would be easier to study the connectedness or otherwise of the set M by using only sets open or closed in X. Fortunately this can be done.

Proposition 7.2 *Let X be a metric space and let $M \subset X$. Then the set M is disconnected if and only if there exist $U \subset X$ and $V \subset X$, open in X and disjoint, such that $M \subset U \cup V$ but neither $M \cap U$ nor $M \cap V$ is empty.*

Proof The condition is sufficient for M to be disconnected since both $M \cap U$ and $M \cap V$ are open relative to M.

Conversely, suppose that M is disconnected. Then $M = A \cup B$ where A and B are open relative to M, disjoint, and neither is empty. For each $x \in A$ there exists $r_x > 0$, such that $B_{r_x}(x) \cap M \subset A$. For each $y \in B$ there exists $s_y > 0$, such that $B_{s_y}(y) \cap M \subset B$. Then for all $x \in A$ and $y \in B$ we have $x \notin B_{s_y}(y)$ and $y \notin B_{r_x}(x)$. By 2.1 exercise 8 this implies $B_{r_x/2}(x) \cap B_{s_y/2}(y) = \emptyset$. Set $U = \bigcup_{x \in A} B_{r_x/2}(x)$ and $V = \bigcup_{y \in B} B_{r_y/2}(y)$. Then U and V are the required sets, open in X. $\quad\square$

A useful paraphrase of the proposition is as follows: the set M, (a subset of X), is connected if and only if, whenever $M \subset A \cup B$, with A and B open in X and $A \cap B = \emptyset$, then either $M \subset A$ or $M \subset B$.

7.1.2 Rules for Connected Sets

Proposition 7.3 *Let M be a connected set in a metric space X. Then every set N that satisfies $M \subset N \subset \overline{M}$ is also connected.*

Proof Suppose that $M \subset N \subset \overline{M}$ but that N is disconnected. Then there exist sets A and B open in X, such that A and B are disjoint, $N \subset A \cup B$, and neither of the intersections $A \cap N$ and $B \cap N$ is empty. Now A and B are open sets that both meet $\overline{M}$ (because they meet N), and therefore they both meet M. We also have $M \subset A \cup B$. But then M is disconnected, contrary to assumption. $\quad\square$

Proposition 7.4 *Let M and N be connected sets in a metric space X, and assume that $M \cap N \neq \emptyset$. Then the set $M \cup N$ is connected.*

Proof Suppose that $M \cup N$ is disconnected. Then there exist disjoint open sets A and B, such that $M \cup N \subset A \cup B$ and neither $A \cap (M \cup N)$ nor $B \cap (M \cup N)$ is empty. This implies in particular that $M \subset A \cup B$ and $N \subset A \cup B$. But since M and N are connected one of the following four cases must hold: (i) $M \subset A$ and $N \subset A$; (ii) $M \subset B$ and $N \subset B$; (iii) $M \subset A$ and $N \subset B$; (iv) $M \subset B$ and $N \subset A$.

Case (i) contradicts the claim that $B \cap (M \cup N) \neq \emptyset$. Case (ii) contradicts the claim that $A \cap (M \cup N) \neq \emptyset$. Cases (iii) and (iv) contradict the assumption that $M \cap N \neq \emptyset$.   □

Proposition 7.5 *Let X be a metric space with the following property: for every pair of points x and y in X there exists a connected set A in X, such that $x \in A$ and $y \in A$. Then X is connected.*

Proof Suppose that $X = U \cup V$ where U and V are disjoint and open. If neither is empty we choose $x \in U$ and $y \in V$. By assumption there then exists a connected set A, such that $x \in A$ and $y \in A$. But then $A \subset U \cup V$ and neither $A \cap U$ nor $A \cap V$ is empty. This contradicts the assumption that A is connected; so we conclude that in fact either U or V is empty. This means that X is connected.   □

Proposition 7.6 *Let X and Y be connected spaces. Then the cartesian product $X \times Y$ is connected.*

Proof Let (x, y) and (x', y') be points in $X \times Y$. The sets $A = \{(x, s) : s \in Y\}$ and $B = \{(t, y') : t \in X\}$ are both connected; the former is isometric with Y and the latter isometric with X. Furthermore they meet, at the point (x, y'). By proposition 7.4 we have that $A \cup B$ is connected, and it contains both (x, y) and (x', y'). By proposition 7.5 we conclude that $X \times Y$ is connected.   □

The proof is pictured in Fig. 7.1.

Fig. 7.1 $X \times Y$ is connected.

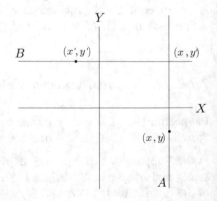

7.1.3 Connected Subsets of $\mathbb{R}$

Proposition 7.7 *The set $M \subset \mathbb{R}$ is connected if and only if it is an interval.*

Proof Let $M \subset \mathbb{R}$ be connected. Let x, y be points in M, with $x < y$, and let $x < z < y$. If $z \notin M$ then

$$M \subset]-\infty, z[\cup]z, \infty[, \quad x \in]-\infty, z[, \quad y \in]z, \infty[.$$

This implies that M is disconnected and contradicts the premise. Therefore $z \in M$; that is to say, if M contains x and y then it contains all points between x and y. It is known from analysis that such a set is an interval, in fact it is one of the intervals $[u, v]$, $[u, v[$, $]u, v]$, $]u, v[$, where $u = \inf M$ and $v = \sup M$ (including the cases $u = -\infty$ and $v = +\infty$).

   Conversely. An open interval $]u, v[$ is homeomorphic to $\mathbb{R}$, and hence connected. By proposition 7.3 all intervals are connected.   □

7.1.4 Exercises

1. Show that the empty set is connected.
2. Let $\mathbf{F}$ be an ordered field with the order topology (section 2.6). Suppose that $\mathbf{F}$ is connected (this means that the only open-closed sets are $\emptyset$ and $\mathbf{F}$). Show that $\mathbf{F}$ is Dedekind complete.

   Hint. Dedekind completeness is the property that every non-empty set A that is bounded above has a least upper bound. Show that the set of all points that are not upper bounds of A is open. Then show that if there is no least upper bound then the set of all upper bounds is also open.

   Note. The result shows that if $\mathbf{F}$ is connected in the order topology then it is a complete ordered field, and hence order isomorphic to $\mathbb{R}$.
3. Let A and B be connected sets in a metric space X, and suppose that $A \cap \overline{B} \neq \emptyset$. Show that $A \cup B$ is connected.
4. Prove the following broad generalisation of proposition 7.4:

   Let $(M_i)_{i \in I}$ be a family of connected sets in X, such that for every partition of the index set $I = I_1 \cup I_2$ into two disjoint non-empty sets we have

   $$\bigcup_{i \in I_1} M_i \cap \bigcup_{j \in I_2} M_j \neq \emptyset.$$

   Then $\bigcup_{i \in I} M_i$ is connected.

5. Let $(M_i)_{i=1}^{\infty}$ be a sequence of connected sets in a metric space X. Suppose that for each i we have $M_i \cap M_{i+1} \neq \emptyset$. Prove that $\bigcup_{i=1}^{\infty} M_i$ is connected.

   Hint. Use the previous exercise.

6. Show that a metric space is connected if and only if every continuous function from X to the two point set $\{0, 1\}$ with the discrete metric, is constant.

7. Let X be a connected metric space and suppose that $X = A \cup B$, where A and B are closed sets such that $A \cap B$ is connected. Show that A and B are both connected. Show that the conclusion remains true if A and B are assumed to be open (instead of closed) but that it need not hold if A is assumed to be open and B closed.

   Hint. A simple (but maybe not obvious) solution can be based on exercise 6, and some rules for building continuous functions by cases (2.3 exercise 13).

8. Let the metric space X be connected and unbounded. Show that for every $a \in X$ and $r > 0$ the sphere $B_r^-(a) \setminus B_r(a)$ is non-empty.

9. Show that a connected metric space that contains at least two distinct points is uncountable, and has cardinality at least **c**.

10. Let X be a connected metric space, let A be a connected set in X and let B be a subset of A^c such that B is open and closed relative to A^c. Show that $A \cup B$ is connected.

    Hint. Use exercise 6 and 2.3 exercise 15.

7.2   Continuous Mappings and Connectedness

In this section we justify the claim made in the preamble to this chapter, by showing that the intermediate value theorem of fundamental analysis follows from the fact that the interval $[a, b]$ is connected.

Proposition 7.8 *Let X and Y be metric spaces and let $f : X \to Y$ be continuous. If X is connected then $f(X)$ is also connected.*

Proof Suppose that $f(X)$ is disconnected. Then we can write $f(X) \subset U \cup V$, where U and V are open in Y, disjoint, and neither $f(X) \cap U$ nor $f(X) \cap V$ is empty. Then $X = f^{-1}(U) \cup f^{-1}(V)$, the sets $f^{-1}(U)$ and $f^{-1}(V)$ are open in X, they are disjoint and neither is empty. This says that X is disconnected, and we have proved the contrapositive of the required implication. □

   The intermediate value theorem follows from this proposition. Let $f : [a, b] \to \mathbb{R}$ be continuous. Then $f([a, b])$ is connected, hence an interval by proposition 7.7, hence contains all points between $f(a)$ and $f(b)$. In particular if $f(a) < 0$ and $f(b) > 0$ then there exists a solution of $f(x) = 0$ between a and b.

7.2.1   Continuous Curves

Let X be a metric space. Let $p \in X$ and $q \in X$. By a *continuous curve* in X with initial point p and final point q we mean a continuous mapping $f : [0, 1] \to X$, such that $f(0) = p$ and $f(1) = q$. We speak loosely here of a curve from p to q.

The range of f is certainly compact and connected. In other respects it could be wildly different form anything that we would naturally like to call a curve in any geometrical sense. For example there exists a surjective continuous mapping $f : [0, 1] \to [0, 1] \times [0, 1]$ (a Peano curve). Worse, there exists a surjective continuous mapping $f : [0, 1] \to H^\infty$ (the range here is the Hilbert cube, see 4.3 exercise 8). This strange phenomenon will be studied in section 7.4. In spite of these examples we shall still use the word 'curve' in this context, though another term would be desirable.

If there exists a continuous curve f from p to q, and also a continuous curve g from q to r, then there exists a continuous curve from p to r. One way to define it is by:

$$h(t) = \begin{cases} f(2t), & 0 \le t < \frac{1}{2} \\ g(2t - 1), & \frac{1}{2} \le t \le 1. \end{cases}$$

7.2.2 Arcwise Connectedness

Using continuous curves we define a highly intuitive type of connectedness, a topological property that turns out to be stronger than connectedness as it was defined in section 7.1.

Definition The space (X, d) is said to be *arcwise connected* if, for every pair of points p and q in X, there exists a continuous curve in X from p to q.

Arcwise connectedness is obviously a topological property. We transfer the predicate to sets in a metric space in the usual way. This means that if X is a metric space and $A \subset X$, then A is an arcwise connected set if, for every pair p and q of points in A, there exists a continuous curve $f : [0, 1] \to X$ such that $f(0) = p$, $f(1) = q$ and $f([0, 1]) \subset A$.

Proposition 7.9 *Arcwise connected spaces are connected.*

Proof Given p and q in an arcwise connected space X there exists a connected set in X (a curve) containing p and q. Hence X is connected by proposition 7.5. □

The converse of this proposition is untrue. There exist connected spaces that are not arcwise connected. However the following result is important and useful for analysis in $\mathbb{R}^n$.

Proposition 7.10 *Let $A \subset \mathbb{R}^n$ be open and connected. Then A is arcwise connected.*

Proof Let $a \in A$. Let M be the set of all points $x \in A$, such that there exists a continuous curve in A joining a to x. Observe that $M \neq \emptyset$, since $a \in M$. We want to show that $M = A$. We shall show that M is both open and closed relative to A.

Let $x \in M$. Since A is open there exists $r > 0$, such that $B_r(x) \subset A$. For each $y \in B_r(x)$ there exists a curve in A joining a to y, since we can first join a to x in A (by the assumption that $x \in M$) and then join x to y along a radius of the ball $B_r(x)$. We conclude that $B_r(x) \subset M$, and therefore M is open in $\mathbb{R}^n$, hence also open relative to A. Note that it does not matter what norm we use in $\mathbb{R}^n$; we only rely on the convexity of the open ball $B_r(x)$.

Let $x \in A \setminus M$. There exists $r > 0$, such that $B_r(x) \subset A$. For no point $y \in B_r(x)$ can a curve exist in A joining a to y, for otherwise there would be a curve in A joining a to x, by first joining a to y, and then y to x along a radius of the ball $B_r(x)$. Therefore $B_r(x) \subset A \setminus M$, and we conclude that $A \setminus M$ is open in $\mathbb{R}^n$, hence open relative to A. So M is closed relative to A.

The upshot is that M is both open and closed relative to A, while, since $a \in M$ it is not empty. Since A is connected we must have $M = A$.   □

The pattern of this proof is instructive and can be useful when we wish to show that all points of a connected metric space X have a given property. We let M be the set of all points in X that have the property, and then establish that M is both open and closed, but non-empty. Then we can conclude that $M = X$. Examples of this argument occur widely in analysis, such as:

a) Ordinary differential equations. The proof that the solution of an initial value problem is unique. See exercise 5.
b) Complex analysis. The proof that two analytic functions in a connected plane set, that coincide at one point together with all their derivatives, are everywhere equal.

7.2.3 Exiting a Set

The notion of boundary ∂A of a set A in a metric space was defined in section 2.2. We recall that

$$\partial A := \overline{A} \setminus A^\circ.$$

It is obvious that ∂A is closed, since $\overline{A}$ is closed and A° open. Points in ∂A are called boundary points of A. We recall that X is expressible as a disjoint union (2.2 exercise 12)

$$X = A^\circ \cup \partial A \cup (A^c)^\circ$$

in which the outer members are both open.

The boundary is unavoidable when we wish to depart a set in a continuous and orderly manner.

Proposition 7.11 *Let X be a metric space, $A \subset X$, $p \in A$ and $q \in A^c$. A continuous curve that joins p and q must meet the boundary ∂A.*

Proof We shall prove a more general conclusion: a connected set B that contains p and q must meet ∂A.

We may assume that $p \in A^\circ$ and $q \in (\overline{A})^c$; otherwise either p or q is itself in ∂A and the conclusion obvious. Assume that the connected set B, which contains p and q, does not meet the boundary ∂A. We deduce a contradiction. We have that $B \subset A^\circ \cup (\overline{A})^c$ and B meets both sets on the right, and they are disjoint and both open. This is impossible since B was supposed to be connected. $\square$

This conclusion agrees well with our intuition about what to expect from a boundary. It was taken for granted by Euclid that a continuous curve (more precisely a circular arc) in a plane that joins a point inside a circle T to a point outside it, must cross T. Euclid gave no axiom that guarantees this, yet it is the basis of the very first proposition of book 1 of 'The Elements', the construction of an equilateral triangle on a given base segment.

7.2.4 Exercises

1. Show that a normed vector space is connected.
2. Show that the unit sphere $S = \{x \in E : \|x\| = 1\}$ in the normed vector space E is connected, with a certain exception that you should specify.
   Hint. Look at the space $E \setminus \{0\}$.
3. An affine function $f : [a, b] \to \mathbb{R}^n$ is a function of the form $f(t) = (1-t)u + tv$ where u and v are given vectors. It parametrises the line segment from u to v. A continuous function $f : [0, 1] \to \mathbb{R}^n$ is called piece-wise affine if there is a partition $0 = a_0 < a_1 < \cdots < a_m = 1$ of its domain, such that f is affine on each subinterval $[a_j, a_{j+1}]$. We also say that f is a polygonal curve joining $f(0)$ and $f(1)$. Affine mappings have appeared in various places in this text, for example, in 2.4, 2.7, 4.5 and 6.3.

   Let A be an open connected subset of $\mathbb{R}^n$. We can obtain a stronger conclusion than that of proposition 7.10.

   a) Prove that any two points in A can be joined by a polygonal curve that lies in A.
   b) Go further and prove that any two points in A can be joined by a polygonal curve that consists of segments each of which is parallel to one of the coordinate axes.

   Hint. You might find the norm $\|u\|_\infty$ useful.

4. Let A be a set in a metric space X.

   a) Show that if A is closed then $(\partial A)^\circ = \emptyset$.
   b) Show that $\partial\partial\partial A = \partial\partial A$.

5. ($\Diamond$) This exercise outlines a proof that the solution to the initial value problem (see section 6.5) is unique.

Let A be open in $\mathbb{R}^2$, let $f : A \to \mathbb{R}$ be continuous and satisfy the following local Lipschitz condition in its second variable: for each $(a, b) \in A$ there exists a neighbourhood U of (a, b) and $K > 0$, such that for all (x, y_1), (x, y_2) in U we have $|f(x, y_1) - f(x, y_2)| \leq K|y_1 - y_2|$. We know by section 6.6 that for each $(a, b) \in A$ there exists an interval $]a - h, a + h[$, on which there exists a unique solution to the initial value problem $y' = f(x, y)$, $y(a) = b$. Now prove the following:

> *If a solution to the initial value problem*
>
> $$y' = f(x, y), \quad y(a) = b$$
>
> *exists on an open interval I (which may be unbounded) containing a, then it is the unique solution with domain I.*

Hint. If two solutions exist on the same interval I, consider the subset of I where they coincide.

6. ($\Diamond$) Let X and Y be Banach spaces. Recall that the space of continuous linear mappings from X to Y is a Banach space $L(X, Y)$ under the operator norm (proposition 3.15). Now let A be a connected subset of $L(X, Y)$ and assume that:

   i) There exists K, such that $\|x\| \leq K\|Tx\|$ for all $x \in X$ and $T \in A$.
   ii) There is at least one surjective element in A.

Show that every element of A is surjective.

Hint. Obviously one wants to show that the set S of all surjective elements of A is open and closed relative to A; or equivalently, S and $A \setminus S$ are both open relative to A. Three things may be found useful to accomplish this. Firstly, show that each T in A is a linear homeomorphism from X on to its range $T(X)$. Secondly, show that $T(X)$ is closed. Thirdly, show that if T_0 is an element of A that is not surjective, and if $y \notin T_0(X)$, then there exists $\delta > 0$, such that $y \notin T(X)$ for all elements T of A that satisfy $\|T - T_0\| < \delta$.

Note. This result can be viewed as an existence theorem for solutions of the equation $T(x) = y$. The inequality in condition (i) is known as an *a priori* bound. It asserts that if a solution exists then it satisfies $\|x\| \leq K\|y\|$. We don't know in advance that a solution exists, but if any do, then they satisfy this bound. This is often implemented in the following way. The set A is the continuous image in $L(X, Y)$ of the unit interval; we have continuous $\phi : [0, 1] \to L(X, Y)$, with $\phi(0) = T_0$ and $\phi(1) = T_1$. The *a priori* bound has the form

$$\|x\| \leq K\|\phi(t)(x)\|, \quad (0 \leq t \leq 1).$$

Now if T_1 is surjective so is T_0, and a solution to $T_0(x) = y$ exists for every y.

7.3 Connected Components

Let (X, d) be a metric space. We shall see that X can be partitioned by maximal connected sets.

Definition Let $a \in X$. The *connected component* of the point a is the union of all connected sets in X that contain a.

Exercise Prove that the connected component of a is connected.

Proposition 7.12 *The connected components of a metric space X form a partition of X, in which each set is the connected component of each of its points.*

In other words, $X = \bigcup_{i \in I} S_i$, where $S_i \cap S_j = \emptyset$ if $i \neq j$, none of the sets S_i is empty, and S_i is the connected component of each point $x \in S_i$.

Proof Define the relation $x \sim y$ to mean that there exists a connected set in X that contains x and y. By proposition 7.4 it is transitive and so is clearly an equivalence relation. Hence X is partitioned into equivalence classes, but obviously each equivalence class is the connected component of each of its points. □

Proposition 7.13 *Connected components are closed.*

Proof Let S be the connected component of the point a. Then S is connected and hence $\overline{S}$ is also connected (proposition 7.3). By the definition of connected component we have $\overline{S} \subset S$ and so S must be closed. □

It is not the case that connected components are always open, though simple examples may make it tempting to think so.

Proposition 7.14 *Let a and b be points in X. Suppose that there exist disjoint open sets A and B, such that $a \in A$, $b \in B$ and $X = A \cup B$. Then a and b have distinct connected components.*

Proof Suppose if possible that a and b have the same connected component M. Then M is connected and $a, b \in M$. But then $M \subset A \cup B$ and M intersects both A and B, which is impossible since M is connected. We conclude that a and b have distinct connected components. □

The converse of this proposition is false as we shall see.

7.3.1 Examples of Connected Components

Verification of these examples will be left to the reader as exercises.

a) $X = [0, 1[\cup]1, 2]$. The connected components are $[0, 1[$ and $]1, 2]$. They are both open (as well as closed).

b) Let X be an open set in $\mathbb{R}$. We have seen (2.2 exercise 28) that X may be expressed as a countable union of pairwise disjoint open intervals $\bigcup_{n=1}^{\infty}]a_n, b_n[$, where a_n may be $-\infty$ and b_n may be $+\infty$. The intervals $]a_n, b_n[$ are the connected components of X. They are all open in X (as well as closed).

c) Cantor's middle thirds set B (viewed as a subspace of $\mathbb{R}$). Each connected component consists of one point only.

A space with the property, that all its connected components consist of one point only, is said to be *totally disconnected*. This is clearly a topological property, even though the property may be verified for Cantor's middle thirds set B by using its embedding into $\mathbb{R}$. Any space homeomorphic to B, such as $2^{\mathbb{N}+}$, is also totally disconnected. We list further examples:

d) $X = \mathbb{Q}$ viewed as a subspace of $\mathbb{R}$. This space is totally disconnected.
e) We describe some subspaces of $\mathbb{R}^2$, called collectively the topologist's comb (Fig. 7.2), that provide counterexamples to some common misconceptions. We let

$$X = \bigcup_{n=1}^{\infty} \left(\{\tfrac{1}{n}\} \times [0, 1] \right)$$

together with some additional points, for example the point $(0, 1)$ and/or $(0, 0)$; or additional segments, such as $[0, 1] \times \{0\}$ or $\{0\} \times [0, 1]$. The connected components depend on the additional points. The analysis is left to the exercises.

The use of connected components is the key to setting up examples of connected spaces that are not arcwise connected. We shall describe one in the following paragraph: a subspace of $\mathbb{R}^2$.

We begin with the spiral curve S given in polar coordinates by

$$r = 1 - e^{-\theta}, \quad (0 \le \theta < \infty).$$

This is a connected set in $\mathbb{R}^2$ since the function $e^{-\theta}$ is continuous. It is easy to see that $\overline{S} = S \cup T$ where T is the unit circle (the reader should check this). By proposition 7.3 the set $X := S \cup T$ is connected.

Fig. 7.2 The topologist's comb

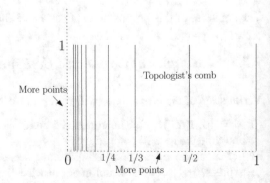

Fig. 7.3 A connected space
that is not arcwise connected

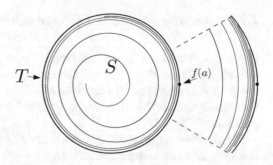

However X is not arcwise connected. There is no continuous curve in X that
joins a point in S to a point in T. At first sight one might think this is so because
the distance along the spiral is infinite. This is incorrect. There is no barrier to a
continuous curve, even one that forms a compact set, having infinite length.[1] We
must find a different argument that is more topological.

Suppose, if possible, that there exists a continuous curve $f : [0, 1] \to X$ with
$p := f(0) \in S$ and $q := f(1) \in T$. We shall derive a contradiction. Consider the
inverse image $f^{-1}(T) \subset [0, 1]$. It is of course closed relative to $[0, 1]$ since T is
closed in X and f is continuous. But $f^{-1}(T)$ is also open relative to $[0, 1]$ as we
shall now see.

Suppose that $a \in f^{-1}(T)$. Let D be a neighbourhood of $f(a)$ in X. It is sufficient
for our purposes to construct D as the intersection of X with a circular sector,
for example with the plane set defined in polar coordinates by the inequalities
$0 < r < 2$, $|\theta - \theta_0| < \pi/4$, where θ_0 is an angular coordinate for $f(a)$. The
neighbourhood D consists of a sequence of arcs of S tending to an arc of T
containing $f(a)$. Each arc, including the limiting one, is a connected component
of D, the latter being viewed as a space in its own right.

Since f is continuous, there exists $\varepsilon > 0$, such that f maps the interval $I_\varepsilon :=$
$[0, 1] \cap]a - \varepsilon, a + \varepsilon[$ (which is a neighbourhood of a relative to $[0, 1]$) into D. Since
intervals are connected, it maps I_ε into one connected component of D, which has
to be the connected component of $f(a)$ in D. Since $f(a)$ lies in T and the connected
component of $f(a)$ in D is an arc of T, it follows that f maps the interval I_ε into T.
Hence $I_\varepsilon \subset f^{-1}(T)$. This shows that $f^{-1}(T)$ is open relative to $[0, 1]$.

Finally, since $[0, 1]$ is connected, $f^{-1}(T)$ both open and closed, and contains a,
we conclude that $f^{-1}(T) = [0, 1]$. This is impossible as f was supposed to join a
point outside T to a point in T.

The argument is illustrated in Fig. 7.3. To the right we see a blown up picture
of the neighbourhood D of $f(a)$, drawing attention to its infinitely many connected
components.

---

[1] An example is the Koch snowflake curve.

7.3.2   Arcwise Connected Components

Given a metric space X we can define the arcwise connected components in a fashion similar to that used for connected components. The arcwise connected component of a point a is the maximal arcwise connected set that contains a. Just as for ordinary connected components, the arcwise connected components constitute a partition of X.

The arcwise connected components can differ from the connected components. However, since an arcwise connected set is connected, it follows that the connected component of a point a includes the arcwise connected component of a. Hence each connected component of X is partitioned by the arcwise connected components of X that it meets (and therefore includes).

7.3.3   Exercises

1. Give arguments for the claims made in items (a–d) in the above list of examples of connected components. Example (e) is the subject of exercise 10.
2. We have seen that the rationals $\mathbb{Q}$ and the Cantor set B are examples of totally disconnected spaces. Show that the following are also totally disconnected:

   a) All discrete spaces.
   b) All countable metric spaces.
   c) All ultrametric spaces (1.2 exercise 1.2.6, 2.1 exercise 9 and 2.2 exercise 18).

3. Let X be a metric space. Show that the connected component of a point a in X is the largest connected set containing a. More precisely, it is the *maximum* element in the set of all connected sets containing a, ordered by inclusion. Among all connected sets in X it is a *maximal* element (compare the preamble to this section).
4. Let X be a metric space, let $p \in X$ and let A be a connected set such that $p \in A$ and A is both open and closed. Show that A is the connected component of p.
5. Suppose that X is a metric space and the number of connected components is finite. Show that each connected component is open (as well as closed).
6. Let X be the subspace of $\mathbb{R}$ consisting of the numbers $1/n$, $n = 1, 2, 3, \ldots$, together with 0. Show that $\{0\}$ is a connected component of X, and moreover it is not open in X.
7. Show that a non-empty, open-closed set A in a metric space X is a union of connected components of X.
8. Let A be a connected component of the disconnected metric space X. Show that there exists an open-closed set U such that $U \neq X$ and $A \subset U$.
9. Let a be a cardinal number. Show that the property of metric spaces expressed by 'the set of all connected components has cardinal number a' is a topological property. In other words the number of connected components is topological.

10. We study the topologist's comb (Fig. 7.2). Let

$$X = \bigcup_{n=1}^{\infty} \left(\{\tfrac{1}{n}\} \times [0, 1] \right),$$

viewed as a subspace of $\mathbb{R}^2$.

a) Show that each segment $\{\tfrac{1}{n}\} \times [0, 1]$ is a connected component of X.

b) Append the segment $[0, 1] \times \{0\}$ to X, producing a new space X_1. Show that X_1 is connected, in fact, arcwise connected.

c) Append the point $p := (0, 1)$ to X_1, producing a new space X_2. Show that X_2 is connected.

d) Remove the segment $[0, 1] \times \{0\}$ from X_2, producing a new space X_3. Show that the set $\{p\}$ is a connected component of X_3. Thus the status of p, whether it is "joined to" its complement or "separated from" it, can change, by removing points "far from" p.

e) Append the points $p := (0, 1)$ and $q := (0, 0)$ to X, producing a new space X_4. Show that the sets $\{p\}$ and $\{q\}$ are connected components of X_4. Show, however, that it is not possible to express X_4 as a union $A \cup B$, where A and B are open relative to X_4, they are disjoint, $p \in A$ and $q \in B$. This provides an example showing that the converse of proposition 7.14 is false.

f) Show that the connected space X_2 defined above is not arcwise connected. Identify its arcwise connected components.

11. Another standard example of a connected but not arcwise connected set is the subspace X of $\mathbb{R}^2$ (the topologist's sine curve) defined by the arc of the graph $y = \sin(1/x)$ for $0 < x < \infty$ together with the point $(0, 0)$. Give arguments for this.

Hint. What does a neighbourhood of $(0, 0)$ in X look like?

12. a) Let A be a connected set in the connected metric space X and let V be a connected component of A^c. Prove that V^c is connected.

Hint. Assume that there exists a set $W \subset V^c$, open and closed relative to V^c, such that $W \neq \emptyset$ and $V^c \setminus W \neq \emptyset$. Obtain a contradiction using 7.1 exercise 10.

b) Let X be a connected metric space having at least two points. Show that there exist disjoint, non-empty, connected subsets M and N such that $X = M \cup N$.

13. Let A be a connected component of the disconnected metric space X. We have seen (exercise 8) that there exists an open-closed set that includes A and is a proper subset of X. The question arises as to whether A is necessarily the intersection of all sets, that are open-closed, and include it. Give a negative answer by building on the topologist's comb.

14. If X is compact the answer to the question raised in the previous exercise is yes. In this exercise we assume that X is compact.

a) Let $(W_i)_{i \in I}$ be a family of closed sets in X with intersection B. Let U be an open set such that $B \subset U$. Show that there is a finite subfamily $(W_{i_k})_{k=1}^n$ such that

$$\bigcap_{k=1}^n W_{i_k} \subset U.$$

Assume that X is disconnected and let C be a connected component of X. Let B be the intersection of all open-closed sets that include C. The next items constitute a proof that $C = B$. At this point we explicitly assume that $C \neq B$, with the aim of deriving a contradiction.

b) Show that B is disconnected.
c) Show that there exist disjoint open sets U and V that both meet B and satisfy $B \subset U \cup V$.
d) Show that there exists an open-closed set W, such that $C \subset W \subset U \cup V$.
e) Show that there exists an open-closed set W', such that either $C \subset W' \subset U$ or $C \subset W' \subset V$.
f) Show that the previous item contradicts the assumed properties of U and V.

In the next two exercises we study *locally connected* spaces. A metric space X is said to be locally connected if, for every $a \in X$, every neighbourhood of a includes a connected neighbourhood of a.

15. a) Show that the example of a space that is connected but not arcwise connected, given at the end of the section 'Examples of connected components', is not locally connected.
    b) Show that the connected space X_2 described in exercise 10(c), is not locally connected.

16. Show that a necessary and sufficient condition for a space X to be locally connected is that, for every open set A in X, the connected components of A are open in X.

    In particular, if X is locally connected the connected components of X itself are open as well as closed. This gives another sufficient condition (to stand beside exercise 14) for an affirmative answer to the question raised in exercise 13.

17. In this exercise we study locally compact spaces and go further in answering the question raised in exercise 13.

    Let X be locally compact and disconnected. Let A be a compact connected component of X. Show that for each $\varepsilon > 0$ there exists an open-closed set W such that

    $$A \subset W \subset N_\varepsilon(A)$$

    where $N_\varepsilon(A)$ is the open ε-neighbourhood of A. Conclude that A is the intersection of all open-closed sets that include it.

Hint. By proposition 4.17 there exists $r > 0$ such that $N_r(A)$ is relatively compact. Apply exercise 14 to the subspace $\overline{N_r(A)}$, of which A is also a connected component.

If A is a non-compact connected component of the locally compact space X then it is not necessarily the case that A is the intersection of all open-closed sets W such that $A \subset W$.

Fig. 7.4 The topologist's rope ladder

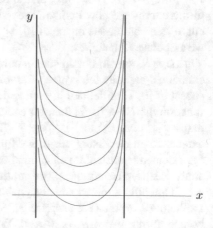

18. We define a subspace X of $\mathbb{R}^2$, illustrated in Fig. 7.4. For each integer $n \in \mathbb{Z}$ let $C_n \subset \mathbb{R}^2$ be the graph of the function

$$y = \frac{1}{x(1-x)} + n, \quad (0 < x < 1).$$

Let L_0 be the line $x = 0$ and L_1 the line $x = 1$. Define

$$X = L_0 \cup L_1 \cup \bigcup_{n \in \mathbb{Z}} C_n.$$

a) Show that X is locally compact.
b) Show that each curve C_n is an open-closed set in X and deduce that each is a connected component of X.
c) Show that the two lines L_0 and L_1 are also connected components of X.
d) Show that any open-closed set that contains a point of L_0 must include the union $L_0 \cup L_1$.
e) Show that the intersection of all open-closed sets that include L_1 is the union $L_0 \cup L_1$.

7.4   ($\Diamond$) Peano

Cantor's conclusion that there exists a bijection from the interval $[0, 1]$ to the square $[0, 1] \times [0, 1]$ stimulated a great deal of speculation in the latter part of the nineteenth century. G. Peano, while trying to construct such a function that might be continuous, found an extraordinary example: a continuous function $F : [0, 1] \to [0, 1] \times [0, 1]$ that is surjective. Many constructions of these space filling curves are now known. They show that there is a problem with defining a curve as a continuous image of $[0, 1]$, if we want it to have the intuitive properties we associate with curves.

A simple way to obtain such a curve, that is capable of being adapted to wider situations, exploits the protean Cantor set. Let B be Cantor's middle thirds set, a subset of $[0, 1]$ as defined in section 1.3. According to 1.3 exercise 2 there exists a function $h : B \to [0, 1]$ that, for each sequence $(a_n)_{n=1}^{\infty}$ in the product space $2^{\mathbb{N}+}$, maps $\sum_{n=1}^{\infty} 2a_n/3^n$ to $\sum_{n=1}^{\infty} a_n/2^n$. This mapping is increasing and surjective. It extends to an increasing, surjective function from $[0, 1]$ to $[0, 1]$ simply by making the extension constant on each maximal open interval in the complement of B. By analysis such a function must be continuous.

We are after a more exotic extension. Recall that the spaces $2^{\mathbb{N}+}$ and $2^{\mathbb{N}+} \times 2^{\mathbb{N}+}$ are homeomorphic (3.2 exercise 3). Since B is homeomorphic to $2^{\mathbb{N}+}$ we have a homeomorphism $j : B \to B \times B$. Write $j = (j_1, j_2)$. Using the function h from the previous paragraph we construct the mapping $k : B \to [0, 1] \times [0, 1]$ by

$$k(t) = \big((h \circ j_1)(t), (h \circ j_2)(t)\big), \quad (t \in B).$$

The mapping k is surjective and continuous.

Now we extend k to the whole interval $[0, 1]$, filling each gap in Cantor's middle thirds set with an affine mapping, whose range must be a subset of $[0, 1] \times [0, 1]$ since the latter is convex. The resulting mapping $F : [0, 1] \to [0, 1] \times [0, 1]$ is continuous.

Exercise Prove that F is continuous. See also exercise 1 below.

A similar argument yields an even weirder conclusion. The space $2^{\mathbb{N}+}$ is also homeomorphic to $(2^{\mathbb{N}+})^{\mathbb{N}+}$ (see 3.2 exercise 4). So B is homeomorphic to $B^{\mathbb{N}+}$. Hence we can build a continuous surjective mapping $k : B \to [0, 1]^{\mathbb{N}+}$, and extend it by bridging the gaps with affine mappings. The result is a continuous surjective mapping $G : [0, 1] \to H^{\infty}$ (the range here is the Hilbert cube, recall).

7.4.1   *Exercises*

1. Let W be a closed, bounded set in $\mathbb{R}$ and let $a = \inf W$, $b = \sup W$. The complement of W is a countable disjoint union of open intervals (2.2 exercise 28). We shall call these intervals, excepting $]-\infty, a[$ and $]b, \infty[$, the

gaps in W. Let the metric space Y be a convex subset of a normed space E. Suppose that $h : W \to Y$ is continuous. Show that h can be extended to a continuous mapping $\tilde{h} : [a, b] \to Y$, by the device of filling in the gaps in W using affine mappings.

Hint. Observe that if the endpoints u and v of the interval $[u, v]$ are in W, then the oscillation of $\tilde{h}$ on $[u, v]$ cannot exceed three times the oscillation of h on $[u, v] \cap W$.

2. Show that a continuous bijection $f : [0, 1] \to [0, 1] \times [0, 1]$ cannot exist.

The continuous, surjective mapping $k : B \to [0, 1]^{\mathbb{N}+}$, where B is Cantor's middle thirds set, that was constructed in this section, can be used to prove the Alexandrov-Hausdorff theorem. This states:

For every non-empty, compact metric space Y there exists a continuous surjective mapping $f : B \to Y$.

By 5.2 exercise 17, a compact metric space Y is homeomorphic to a subspace of $[0, 1]^{\mathbb{N}+}$. So we can assume that Y is a subspace of $[0, 1]^{\mathbb{N}+}$. Let $W = k^{-1}(Y)$. Then W is a closed set in B and the restriction of k to W maps it on to Y. It is therefore enough to prove that a continuous, surjective mapping exists from B to W. This can then be composed with $k|_W$.

3. Prove that if W is a closed subset of B then there exists a continuous and surjective mapping $g : B \to W$. You can use the following steps:

   a) The result to be proved is clearly a topological property of B. We can therefore replace B by a space homeomorphic to it, yet another incarnation of the Cantor set. Let B' be the set of all real numbers in the interval $[0, 1]$, that can be written in base 5 as $x = (0.a_1a_2a_3\ldots)_5$, where each digit is 0 or 4. Show that B' is homeomorphic to B.

   b) Let V be a closed subset of B'. Let $x \in B'$. Show that there exists a unique $y \in V$ such that $|x - y| = d(x, V)$.

      Hint. Can you have three distinct points in B' in arithmetic progression?

   c) Define $F : B' \to V$ by setting $F(x) = y$, where y is the unique $y \in V$ such that $|x - y| = d(x, V)$. Show that F is continuous and surjective.

7.4.2 Pointers to Further Study

→   Space filling curves (Peano curve, Hilbert curve)
→   Computer graphics and data structuring
→   Hahn-Mazurkiewicz theorem

Afterword

Most books on higher analysis make heavy use of metric space concepts, and it is impossible to list them here. However, to make amends, we can name a few, all of them classics, which should be easily accessible to the reader of the current volume.

1. W. Rudin. Principles of mathematical analysis.

Covers fundamental analysis and a chunk of multivariate analysis, using metric spaces from the start. It remains popular even 70 years after it first appeared. Exuding the author's characteristic enthusiasm, it pours out definitions and theorems in quick succession, though for the uninitiated the motivation may seem somewhat skimpy.

2. W. Rudin. Real and complex analysis.

Continues in the same style. Covers measure and integration, Fourier analysis and complex analysis, with a somewhat functional analytic viewpoint. Many theorems are stated in the context of locally compact Hausdorff spaces. In practical applications, and in particular in multivariate analysis, these will nearly always be metric spaces.

3. J. Dieudonné. Foundations of modern analysis.

Has an undeserved reputation for being hard, possibly due to the author's dry style (no pictures!) and a certain haughtiness towards alternative ways of doing things. Nevertheless, it is well worth reading. Covers differential calculus in normed spaces and a small chunk of functional analysis, with the necessary metric space theory, which is similar in scope to the present volume.

© The Author(s), under exclusive license to Springer Nature Switzerland AG 2022
R. Magnus, *Metric Spaces*, Springer Undergraduate Mathematics Series,
https://doi.org/10.1007/978-3-030-94946-4

4. E. Kreyszig. Introductory functional analysis with applications.

A very readable introduction to functional analysis. It should appeal to a wide range of interests on the spectrum from "pure" to "applied" mathematics. It is nevertheless fully rigorous and presents the usual big theorems (Hahn-Banach, Banach-Steinhaus, etc.)

   We list next some books on topology and set theory.

5. I. Kaplansky. Set theory and metric spaces.

Written in the author's attractive and laid back style this can offer the reader all the set theory (and more) that is needed for the current volume. The chapters on metric spaces are viewed by the author as a stepping stone to general topology, rather than as a tool for analysis. For this reason, connected spaces are omitted entirely and product spaces scantily treated.

6. J. Munkres. Topology.

This is a standard work on general topology, written in a readable and friendly style, with metric spaces included as a fairly minor special case.

Index

A
Affine mapping, 68
Alexandrov-Hausdorff theorem, 237
A priori bound, 228
Archimedean axiom, 187
Arcwise connected
    components, 232
    space, 225
Ascoli's theorem, 128, 156

B
Baire
    category theorem, 190
    space, 190, 191
Banach-Mazurkiewicz theorem, 199
Banach space, 90
Banach-Steinhaus theorem, 192
Base
    for neighbourhoods, 144
    for topology, 172
Bernstein polynomial, 177
Bézier curves, 181
Binary tree, 19
Bolzano-Weierstrass theorem, 126
Borel
    measure, 77
    set, 63
Bounded set, 32
Bourbaki, N., 167

C
Cantor, G., 16, 188, 191, 199, 236
Cantor set, 16–18, 47, 93, 192, 197, 198, 230, 236, 237

Cantor's nested intersection theorem, 187, 189
Category, 188
Cauchy-Schwarz inequality, 11
Cauchy sequence, 33
Cauchy's principle, 55
Clarkson's inequalities, 26
Closed
    ball, 29
    relative to a set, 45
    set, 39
Closure, 42
Compact
    set, 132
    space, 131, 137, 149
Complete
    metric space, 34
    ordered field, 34, 223
Completeness
    of ℓ^p, 88
    of $A(D)$, 148
    of $C(\mathbb{R})$, 104
    of $C^p(\mathbb{R})$, 104
    of $C_B(M)$, 101
    of $\mathbb{R}^n$, 88
    of $\mathcal{O}(A)$, 149
Completion (of a metric space), 216–218
Connected
    component, 229
    set, 221
    space, 220
Continuous
    curve, 224
    linear mapping, 63–65
    mapping, 52–54

© The Author(s), under exclusive license to Springer Nature Switzerland AG 2022
R. Magnus, *Metric Spaces*, Springer Undergraduate Mathematics Series,
https://doi.org/10.1007/978-3-030-94946-4

Contraction, 204
    mapping principle, 203
    strict, 204
Convergence
    pointwise, 99
    uniform, 99
Convex hull, 49

D
Dense set, 69, 155, 163–165
Diameter (of a set), 3
Dini's theorem, 136
Discrete
    metric, 35
    space, 71
    topology, 75
Distance (from a point to a set), 3
Distribution, 187
Dyadic metric, 9, 14, 35

E
Equicontinuous
    family of functions, 155
    set of functions, 154
Equivalent metrics, 72
Euclid, 11, 227
Euler, L., 158

F
F_σ-set, 63
Finite intersection property, 136
First category set, 188, 189, 191, 194, 199, 201
First countable space, 173
Fixed point, 203
Fixed point theorem
    Banach's, 203–207, 209, 210
    Brouwer's, 204
    Schauder's, 204
Fourier transform, 170
Frechet derivative, 212
Fréchet, M.R., 212
Fredholm integral equation, 115
Function space, 22
Fundamental Theorem of Algebra, 130

G
G_δ-set, 63
Gauss, C.F., 126
Generating set
    for a σ-algebra, 80

for a topology, 79
Genericity
    probabilistic, 193, 197
    topological, 194, 197, 201
Graph (as metric space), 9

H
Hausdorff, F., 63
    metric, 10, 15, 48
    space, 76
Heine-Borel theorem, 140, 150
Hilbert cube, 128, 143, 154, 175, 225, 236
Hölder's inequality, 12
    for integrals, 24
Homeomorphism, 70

I
Indiscrete topology, 75
Integral
    Lebesgue, 168, 170, 187
    regulated, 168
    Riemann, 168, 187
    vector valued, 165–167
Interior, 42
Isolated point, 47
Isometry, 82, 164, 216

K
Kantorovich, L.V., 210

L
Lattice, 129
Lebesgue
    measure, 20, 77, 197
    number, 136
Limit
    of a function, 55
    of a sequence, 4, 30
Limit point (of a set), 18, 43
Limit point compactness, 137
Lindelöf space, 175
Liouville, J., 199
Lipschitz mapping, 58
Locally compact
    set, 142
    space, 142, 234
Locally connected space, 234
Lowest uncountable cardinal, 81

M
Mazur-Ulam theorem, 82
Measurable space, 77
Measure, 77
    space, 77
Metric
    definition of, 3
    symmetry axiom, 3
    triangle inequality, 3
Metric space (definition of), 3
Minkowski's inequality, 12
    for integrals, 24
Montel's theorem, 162

N
Nash, J., 204
Neighbourhood
    of a point, 32
    of a set, 51
Neumann, C., 113
Neumann series, 113
Newton
    approximation method, 210
    iteration, 212
Newton-Kantorovich theorem, 214
Norm (definition of), 7
Normal space, 76
Normed
    algebra, 128
    space (definition of), 7
Nowhere
    dense subset, 188
    differentiable function, 199

O
Open
    ball, 29
    covering, 131
    relative to a set, 45
    set, 38
Operator norm, 66, 67
Order topology, 78, 149
Oscillation, 56

P
Peano, G., 236
Peano's existence theorem, 157–160
Picard-Lindelöf theorem, 206, 207, 209
Product
    metric, 59
    space, 90–93, 97
Pseudometric, 9

R
Regulated function, 147, 167
Relatively compact set, 141
Residual set, 191, 194, 196, 201
Riesz's lemma, 152
Rising sun lemma, 60

S
σ-algebra, 76
Second axiom of countability, *see* Second
                countable space
Second category set, 188–191, 201
Second countable space, 172
Seminorm, 16
Separable space, 71, 171–173
Separation (of two sets), 129
Sequence space, 22
    ℓ^p, 21
    $\mathbb{R}^{\mathbb{N}_+}$, 21, 25
Sequentially compact
    set, 126–128
    space, 71, 125, 137
Series
    absolute convergence of, 102
    conditional convergence of, 108
    in a normed space, 101
    rearrangements of, 108
    unconditional convergence of, 108
Space
    $A(D)$, 148
    $B(M)$, 22
    $C(M)$, 99
    $C(\mathbb{R})$, 103
    $C^p(\mathbb{R})$, 103
    $C_0(M)$, 129
    $C_0(\mathbb{R})$, 49
    $C_B(M)$, 99
    $L(X, Y)$ (of operators), 66
    $\mathbb{R}^{\mathbb{N}_+}$, 25
    $\mathcal{O}(A)$, 148
    c, 90
    c_0, 90
    of continuous functions on $[a, b]$, 23
    of differentiable functions
            on $[a, b]$, 25
    of integrable functions, 23
Stone-Weierstrass theorem, 182–184
Subbase (for a topology), 80, 97
Subspace (of a metric space), 8, 45

T
Tietze's extension theorem, 118
Topological

embedding, 70
property, 71
space, 75
Topologist's
comb, 230, 233
rope ladder, 235
sine curve, 233
Topology, 75
Totally
bounded space, 137, 138
disconnected space, 230
Transcendental number, 199
Tychonov's theorem, 149

U
Ultrametric space, 14, 35, 49, 96
Uniform boundedness principle, 192
Uniformly
continuous function, 130, 135
convergent function sequence, 23

convex (normed space), 27
Upper semi-continuous, 63
Urysohn's lemma, 118

V
Vitali covering lemma, 36
Volterra integral equation, 116
Von Neumann, J., 204

W
Weak topology (on a vector space), 81
Weierstrass approximation theorem, 176–180
Weierstrass, K., 199
Weierstrass M-test, 103

Y
Young's inequality, 12

Printed in the United States
by Baker & Taylor Publisher Services